The First Three Minutes

THE FIRST THREE MINUTES

A Modern View of the
Origin of the Universe

STEVEN WEINBERG

ANDRE DEUTSCH

To my parents

First published 1977 by
André Deutsch Limited
105 Great Russell Street London WC1

Printed in Great Britain by
Ebenezer Baylis and Son Ltd
The Trinity Press, Worcester, and London

ISBN 0 233 96906 3

Contents

Contents

vi

Preface

This book grew out of a talk I gave at the dedication of the Undergraduate Science Center at Harvard in November 1973. Erwin Glikes, president and publisher of Basic Books, heard of this talk from a mutual friend, Daniel Bell, and urged me to turn it into a book.

At first I was not enthusiastic about the idea. Although I have done small bits of research in cosmology from time to time, my work has been much more concerned with the physics of the very small, the theory of elementary particles. Also, elementary particle physics has been extraordinarily lively in the last few years, and I had been spending too much time away from it, writing nontechnical articles for various magazines. I wanted very much to return full time to my natural habitat, the *Physical Review*.

However, I found that I could not stop thinking about the idea of a book on the early universe. What could be more interesting than the problem of Genesis? Also, it is in the early universe, especially the first hundredth of a second, that the problems of the theory of elementary particles come together with the problems of cosmology. Above all, this is a good time to write about the early universe. In just the last decade a detailed theory of the course of events in the early universe has become widely accepted as a "standard model."

It is a remarkable thing to be able to say just what the universe was like at the end of the first second or the first minute or the first year. To a physicist, the exhilarating thing is to be

Preface

able to work things out numerically, to be able to say that at such and such a time the temperature and density and chemical composition of the universe had such and such values. True, we are not absolutely certain about all this, but it is exciting that we are now able to speak of such things with any confidence at all. It was this excitement that I wanted to convey to the reader.

I had better say for what reader this book is intended. I have written for one who is willing to puzzle through some detailed arguments, but who is not at home in either mathematics or physics. Although I must introduce some fairly complicated scientific ideas, no mathematics is used in the body of the book beyond arithmetic, and little or no knowledge of physics or astronomy is assumed in advance. I have tried to be careful to define scientific terms when they are first used, and in addition I have supplied a glossary of physical and astronomical terms (p. 157). Wherever possible, I have also written numbers like "a hundred thousand million" in English, rather than use the more convenient scientific notation: 10^{11}.

However, this does not mean that I have tried to write an easy book. When a lawyer writes for the general public, he assumes that they do not know Law French or the Rule Against Perpetuities, but he does not think the worse of them for it, and he does not condescend to them. I want to return the compliment: I picture the reader as a smart old attorney who does not speak *my* language, but who expects nonetheless to hear some convincing arguments before he makes up his mind.

For the reader who does want to see some of the calculations that underlie the arguments of this book, I have prepared "A Mathematical Supplement," which follows the body of the book (p. 166). The level of mathematics used here would make these notes accessible to anyone with an undergraduate concentration in any physical science or mathematics. Fortunately, the most important calculations in cosmology *are* rather simple; it is only here and there that the finer points of general relativity or nuclear physics come into play. Readers who want to pursue this subject on a more technical level will find several advanced treatises (including my own) listed under "Suggestions for Further Reading" (p. 177).

I should also make clear what subject I intended this book to cover. It is definitely not a book about all aspects of cosmology. There is a "classic" part of the subject, which has to do mostly with the large-scale structure of the present universe: the debate over the extragalactic nature of the spiral nebulae; the discovery of the red shifts of distant galaxies and their dependence on distance; the general relativistic cosmological models of Einstein, de Sitter, Lemaitre, and Friedmann; and so on. This part of cosmology has been described very well in a number of distinguished books, and I did not intend to give another full account of it here. The present book is concerned with the early universe, and in particular with the new understanding of the early universe that has grown out of the discovery of the cosmic microwave radiation background in 1965.

Of course, the theory of the expansion of the universe is an essential ingredient in our present view of the early universe, so I have been compelled in Chapter II to provide a brief introduction to the more "classic" aspects of cosmology. I believe that this chapter should provide an adequate background, even for the reader completely unfamiliar with cosmology, to understand the recent developments in the theory of the early universe with which the rest of the book is concerned. However, the reader who wants a thorough introduction to the older parts of cosmology is urged to consult the books listed under "Suggestions for Further Reading."

On the other hand, I have not been able to find any coher-

Preface

ent historical account of the recent developments in cosmology. I have therefore been obliged to do a little digging myself, particularly with regard to the fascinating question of why there was no search for the cosmic microwave radiation background long before 1965. (This is discussed in Chapter VI.) This is not to say that I regard this book as a definitive history of these developments—I have far too much respect for the effort and attention to detail needed in the history of science to have any illusions on that score. Rather, I would be happy if a real historian of science would use this book as a starting point, and write an adequate history of the last thirty years of cosmological research.

I am extremely grateful to Erwin Glikes and Farrell Phillips of Basic Books for their valuable suggestions in preparing this manuscript for publication. I have also been helped more than I can say in writing this book by the kind advice of my colleagues in physics and astronomy. For taking the trouble to read and comment on portions of the book, I wish especially to thank Ralph Alpher, Bernard Burke, Robert Dicke, George Field, Gary Feinberg, William Fowler, Robert Herman, Fred Hoyle, Jim Peebles, Arno Penzias, Bill Press, Ed Purcell, and Robert Wagoner. My thanks are also due to Isaac Asimov, I. Bernard Cohen, Martha Liller, and Philip Morrison for information on various special topics. I am particularly grateful to Nigel Calder for reading through the whole of the first draft, and for his perceptive comments. I cannot hope that this book is now entirely free of errors and obscurities, but I am certain that it is a good deal clearer and more accurate than it could have been without all the generous assistance I have been fortunate enough to receive.

STEVEN WEINBERG

Cambridge, Massachusetts
July 1976

The First Three Minutes

INTRODUCTION: THE GIANT AND THE COW

THE ORIGIN of the universe is explained in the *Younger Edda*, a collection of Norse myths compiled around 1220 by the Icelandic magnate Snorri Sturleson. In the beginning, says the *Edda*, there was nothing at all. "Earth was not found, nor Heaven above, a Yawning-gap there was, but grass nowhere." To the north and south of nothing lay regions of frost and fire, Niflheim and Muspelheim. The heat from Muspelheim melted some of the frost from Niflheim, and from the liquid drops there grew a giant, Ymer. What did Ymer eat? It seems there was also a cow, Audhumla. And what did *she* eat? Well, there was also some salt. And so on.

I must not offend religious sensibilities, even Viking religious sensibilities, but I think it is fair to say that this is not a very satisfying picture of the origin of the universe. Even leaving aside all objections to hearsay evidence, the story raises as many problems as it answers, and each answer requires a new complication in the initial conditions.

We are not able merely to smile at the *Edda*, and forswear all cosmogonical speculation—the urge to trace the history of the universe back to its beginnings is irresistible. From the start of modern science in the sixteenth and seventeenth centuries, physicists and astronomers have returned again and again to the problem of the origin of the universe.

However, an aura of the disreputable always surrounded such research. I remember that during the time that I was a student and then began my own research (on other problems) in the 1950s, the study of the early universe was widely regarded as not the sort of thing to which a respectable scientist would devote his time. Nor was this judgment unreasonable. Throughout most of the history of modern physics and astronomy, there simply has not existed an adequate observational and theoretical foundation on which to build a history of the early universe.

Now, in just the past decade, all this has changed. A theory of the early universe has become so widely accepted that astronomers often call it "the standard model." It is more or less the same as what is sometimes called the "big bang" theory, but supplemented with a much more specific recipe for the contents of the universe. This theory of the early universe is the subject of this book.

To help see where we are going, it may be useful to start with a summary of the history of the early universe, as presently understood in the standard model. This is only a brief run-through—succeeding chapters will explain the details of this history, and our reasons for believing any of it.

In the beginning there was an explosion. Not an explosion like those familiar on earth, starting from a definite center and spreading out to engulf more and more of the circumambient air, but an explosion which occurred simultaneously everywhere, filling all space from the beginning, with every particle of matter rushing apart from every other particle. "All space" in this context may mean either all of an infinite universe, or all of a finite universe which curves back on itself like the surface of a sphere. Neither possibility is easy to comprehend, but this will not get in our way; it matters hardly at all in the early universe whether space is finite or infinite.

At about one-hundredth of a second, the earliest time about which we can speak with any confidence, the temperature of the universe was about a hundred thousand million (10^{11}) degrees Centigrade. This is much hotter than in the center of even the hottest star, so hot, in fact, that none of the components of ordinary matter, molecules, or atoms, or even the nuclei of atoms, could have held together. Instead, the matter rushing apart in this explosion consisted of various types of the so-called elementary particles, which are the subject of modern high-energy nuclear physics.

We will encounter these particles again and again in this book—for the present it will be enough to name the ones that were most abundant in the early universe, and leave more detailed explanations for Chapters III and IV. One type of particle that was present in large numbers is the electron, the negatively charged particle that flows through wires in electric currents and makes up the outer parts of all atoms and molecules in the present universe. Another type of particle that was abundant at early times is the positron, a positively charged particle with precisely the same mass as the electron. In the present universe positrons are found only in high-energy laboratories, in some kinds of radioactivity, and in violent astronomical phenomena like cosmic rays and supernovas, but in the early universe the number of positrons was almost exactly equal to the number of electrons. In addition to electrons and positrons, there were roughly similar numbers of various kinds of neutrinos, ghostly particles with no mass or electric charge whatever. Finally, the universe was filled with light. This does not have to be treated separately from the particles—the quantum theory tells us that light consists of particles of zero mass and zero electrical charge known as photons. (Each time an atom in the filament of a light bulb changes from a state of higher energy to one of lower energy, one photon is emitted. There are so many photons coming out of a light bulb that they seem to blend together in a continuous stream of light, but a photoelectric cell can count individual photons, one by one.) Every photon carries a definite amount of energy and momentum depending on the wavelength of the light. To describe the light that filled the early universe, we can say that the number and the average energy of the photons was about the same as for electrons or positrons or neutrinos.

These particles—electrons, positrons, neutrinos, photons— were continually being created out of pure energy, and then after short lives being annihilated again. Their number therefore was not preordained, but fixed instead by a balance between processes of creation and annihilation. From this balance we can infer that the density of this cosmic soup at a temperature of a hundred thousand million degrees was about four thousand million (4×10^9) times that of water. There was also a small contamination of heavier particles, protons and neutrons, which in the present world form the constituents of atomic nuclei. (Protons are positively charged; neutrons are slightly heavier and electrically neutral.) The proportions were roughly one proton and one neutron for every thousand million electrons or positrons or neutrinos or photons. This number—a thousand million photons per nuclear particle—is the crucial quantity that had to be taken from observation in order to work out the standard model of the universe. The discovery of the cosmic radiation background discussed in Chapter III was in effect a measurement of this number.

As the explosion continued the temperature dropped, reaching thirty thousand million (3×10^{10}) degrees Centigrade after about one-tenth of a second; ten thousand million degrees after about one second; and three thousand million degrees after about fourteen seconds. This was cool enough so that the electrons and positrons began to annihilate faster than they could be recreated out of the photons and neutrinos. The energy released in this annihilation of matter temporarily slowed the rate at which the universe cooled, but the temperature continued to drop, finally reaching one thousand million degrees at the end of the first three minutes. It was then cool enough for the protons and neutrons to begin to form into complex nuclei, starting with the nucleus of heavy hydrogen (or deuterium), which consists of one proton and one neutron. The density was still high enough (a little less than that of water) so that these light nuclei were able rapidly to assemble themselves into the most stable light nucleus, that of helium, consisting of two protons and two neutrons.

At the end of the first three minutes the contents of the universe were mostly in the form of light, neutrinos, and antineutrinos. There was still a small amount of nuclear material, now consisting of about 73 percent hydrogen and 27 percent

7

helium, and an equally small number of electrons left over from the era of electron-positron annihilation. This matter continued to rush apart, becoming steadily cooler and less dense. Much later, after a few hundred thousand years, it would become cool enough for electrons to join with nuclei to form atoms of hydrogen and helium. The resulting gas would begin under the influence of gravitation to form clumps, which would ultimately condense to form the galaxies and stars of the present universe. However, the ingredients with which the stars would begin their life would be just those prepared in the first three minutes.

The standard model sketched above is not the most satisfying theory imaginable of the origin of the universe. Just as in the *Younger Edda,* there is an embarrassing vagueness about the very beginning, the first hundredth of a second or so. Also, there is the unwelcome necessity of fixing initial conditions, especially the initial thousand-million-to-one ratio of photons to nuclear particles. We would prefer a greater sense of logical inevitability in the theory.

For example, one alternative theory that seems philosophically far more attractive is the so-called steady-state model. In this theory, proposed in the late 1940s by Herman Bondi, Thomas Gold, and (in a somewhat different formulation) Fred Hoyle, the universe has always been just about the same as it is now. As it expands, new matter is continually created to fill up the gaps between the galaxies. Potentially, all questions about why the universe is the way it is can be answered in this theory by showing that it is the way it is because that is the only way it can stay the same. The problem of the early universe is banished; there was no early universe.

How then did we come to the "standard model?" And how has it supplanted other theories, like the steady-state model? It is a tribute to the essential objectivity of modern astrophysics that this consensus has been brought about, not by shifts in philosophical preference or by the influence of astrophysical mandarins, but by the pressure of empirical data.

The next two chapters will describe the two great clues, furnished by astronomical observation, which have led us to the standard model—the discoveries of the recession of distant galaxies and of a weak radio static filling the universe. This is a rich story for the historian of science, filled with false starts, missed opportunities, theoretical preconceptions, and the play of personalities.

Following this survey of observational cosmology, I will try to put the pieces of data together to make a coherent picture of physical conditions in the early universe. This will put us in a position to go back over the first three minutes in greater detail. A cinematic treatment seems appropriate: frame by frame, we will watch the universe expand and cool and cook. We will also try to look a little way into an era that is still clothed in mystery—the first hundredth of a second, and what went before.

Can we really be sure of the standard model? Will new discoveries overthrow it and replace the present standard model with some other cosmogony, or even revive the steady-state model? Perhaps. I cannot deny a feeling of unreality in writing about the first three minutes as if we really know what we are talking about.

However, even if it is eventually supplanted, the standard model will have played a role of great value in the history of cosmology. It is now respectable (though only in the last decade or so) to test theoretical ideas in physics or astrophysics by working out their consequences in the context of the standard model. It is also common practice to use the standard model as a theoretical basis for justifying programs of astronomical observation. Thus, the standard model provides an essential common language which allows theorists and observers to appreciate what each other are doing. If some day the standard model is replaced by a better theory, it will probably be because of observations or calculations that drew their motivation from the standard model.

In the last chapter I will say a bit about the future of the universe. It may go on expanding forever, getting colder, emptier, and deader. Alternatively, it may recontract, breaking up the galaxies and stars and atoms and atomic nuclei back into their constituents. All the problems we face in understanding the first three minutes would then arise again in predicting the course of events in the last three minutes.

II

THE EXPANSION OF THE UNIVERSE

A LOOK at the night sky gives a powerful impression of a changeless universe. True, clouds drift across the moon, the sky rotates around the polar star, and over longer times the moon itself waxes and wanes and the moon and planets move against the background of stars. But we know that these are merely local phenomena caused by motions within our solar system. Beyond the planets, the stars seem motionless.

Of course, the stars do move, at speeds ranging up to a few hundred kilometers per second, so in a year a fast star might travel ten thousand million kilometers or so. This is a thousand times less than the distance to even the closest stars, so their apparent position in the sky changes very slowly. (For instance, the relatively fast star known as Barnard's star is at a distance of about 56 million million kilometers; it moves across the line of sight at about 89 kilometers per second or 2.8 thousand million kilometers per year, and in consequence its apparent position shifts in one year by an angle of 0.0029 degrees.) Astronomers call the shift in the apparent position of nearby stars in the sky a "proper motion." The apparent positions in the sky of the more distant stars change so slowly that their proper motion cannot be detected with even the most patient observation.

We are going to see here that this impression of changelessness is illusory. The observations that we will discuss in this chapter reveal that the universe is in a state of violent explosion, in which the great islands of stars known as galaxies are rushing apart at speeds approaching the speed of light. Further, we can extrapolate this explosion backward in time and conclude that all the galaxies must have been much closer at the same time in the past—so close, in fact, that neither galaxies nor stars nor even atoms or atomic nuclei could have had a separate existence. This is the era we call "the early universe," which serves as the subject of this book.

Our knowledge of the expansion of the universe rests entirely on the fact that astronomers are able to measure the motion of a luminous body in a direction directly *along* the line of sight much more accurately than they can measure its motion at right angles to the line of sight. The technique makes use of a familiar property of any sort of wave motion, known as the Doppler effect. When we observe a sound or light wave from a source at rest, the time between the arrival of wave crests at our instruments is the same as the time between crests as they leave the source. On the other hand, if the source is moving away from us, the time between arrivals of successive wave crests is increased over the time between their departures from the source, because each crest has a little farther to go on its journey to us than the crest before. The time between crests is just the wavelength divided by the speed of the wave, so a wave sent out by a source moving away from us will appear to have a *longer* wavelength than if the source were at rest. (Specifically, the fractional increase in the wavelength is given by the ratio of the speed of the wave source to the speed of the wave itself, as shown in mathematical note 1, p. 166.) Similarly, if the source is moving toward us, the time between arrivals of wave crests is decreased because each successive crest has a shorter distance to go, and the wave appears to have a *shorter* wavelength. It is just as if a traveling salesman were to send a letter home regularly once a week during his travels: while he is traveling away from home, each successive letter will have a little farther to go than the one before, so his letters will arrive a little more than a week apart; on the homeward leg of his journey, each successive letter will have a shorter distance to travel, so they will arrive more frequently than once a week.

It is easy these days to observe the Doppler effect on sound waves—just go out to the edge of a highway and notice that the engine of a fast automobile sounds higher pitched (i.e., a shorter wavelength) when the auto is approaching than when it is going away. The effect was apparently first pointed out for both light and sound waves by Johann Christian Doppler, professor of mathematics at the Realschule in Prague, in 1842. The Doppler effect for sound waves was tested by the Dutch meteorologist Christopher Heinrich Dietrich Buys-Ballot in an endearing experiment in 1845—as a moving source of sound he used an orchestra of trumpeters standing in an open car of a railroad train, whizzing through the Dutch coun-tryside near Utrecht.

Doppler thought that his effect might explain the different

colors of stars. The light from stars that happen to be moving away from the earth would be shifted toward longer wavelengths, and since red light has a wavelength longer than the average wavelength for visible light, such a star might appear redder than average. Similarly, light from stars that happen to be moving toward the earth would be shifted toward shorter wavelengths, so the star might appear unusually blue. It was soon pointed out by Buys-Ballot and others that the Doppler effect has essentially nothing to do with the color of a star—it is true that the blue light from a receding star is shifted toward the red, but at the same time some of the star's normally invisible ultraviolet light is shifted into the blue part of the visible spectrum, so the overall color does not change. Stars have different colors chiefly because they have different surface temperatures.

However, the Doppler effect did begin to be of enormous importance to astronomy in 1868, when it was applied to the study of individual spectral *lines*. It had been discovered years earlier, by the Munich optician Joseph Frauenhofer in 1814–1815, that when light from the sun is allowed to pass through a slit and then through a glass prism, the resulting spectrum of colors is crossed with hundreds of dark lines, each one an image of the slit. (A few of these lines had been noticed even earlier, by William Hyde Wollaston in 1802, but were not carefully studied at that time.) The dark lines were always found at the same colors, each corresponding to a definite wavelength of light. The same dark spectral lines were also found by Frauenhofer in the same positions in the spectrum of the moon and the brighter stars. It was soon realized that these dark lines are produced by the selective absorption of light of certain definite wavelengths, as the light passes from the hot surface of a star through its cooler outer atmosphere. Each line is due to the absorption of light by a specific chemical element, so it became possible to determine that the elements on the sun, such as sodium, iron, magnesium, calcium, and chromium, are the same as those found on earth. (Today we know that the wavelengths of the dark lines are just those for which a photon of that wavelength would have precisely the right energy to raise the atom from its state of lowest energy to one of its excited states.)

In 1868 Sir William Huggins was able to show that the dark lines in the spectra of some of the brighter stars are shifted slightly to the red or the blue from their normal position in the spectrum of the sun. He correctly interpreted this as a Doppler shift, due to the motion of the star away from or toward the earth. For instance, the wavelength of every dark line in the spectrum of the star Capella is longer than the wavelength of the corresponding dark line in the spectrum of the sun by 0.01 percent; this shift to the red indicates that Capella is receding from us at 0.01 percent of the speed of light, or 30 kilometers per second. The Doppler effect was used in the following decades to discover the velocities of solar prominences, of double stars, and of the rings of Saturn.

The measurement of velocities by the observation of Doppler shifts is an intrinsically accurate technique, because the wavelengths of spectral lines can be measured with very great precision; it is not unusual to find wavelengths given in tables to eight significant figures. Also, the technique preserves its accuracy whatever the distance of the light source, provided only that there is enough light to pick out spectral lines against the radiation of the night sky.

It is through use of the Doppler effect that we know the typical values of stellar velocities referred to at the beginning of this chapter. The Doppler effect also gives us a clue to the distances of nearby stars; if we guess something about a star's direction of motion, then the Doppler shift gives us its speed across as well as along our line of sight, so measurement of the star's apparent motion across the celestial sphere tells us how far away it is. But the Doppler effect began to give results of cosmological importance only when astronomers began to study the spectra of objects at a much greater distance than the visible stars. I will have to say a bit about the discovery of those objects and then come back to the Doppler effect.

We started this chapter with a look at the night sky. In addition to the moon, planets, and stars, there are two other visible objects, of greater cosmological importance, that I might have mentioned.

One of these is so conspicuous and brilliant that it is sometimes visible even through the haze of a city's night sky. It is the band of lights stretching in a great circle across the celestial sphere, and known from ancient times as the Milky Way. In 1750 the English instrument maker Thomas Wright published a remarkable book, *Original Theory or New Hypothesis of the Universe*, in which he suggested that the stars lie in a flat slab, a "grindstone," of finite thickness but extending to great distances in all directions in the plane of the slab. The solar system lies within the slab, so naturally we see much more light when we look out from earth along the plane of the slab than when we look in any other direction. This is what we see as the Milky Way.

Wright's theory has long since been confirmed. It is now thought that the Milky Way consists of a flat disk of stars, with a diameter of 80,000 light years and a thickness of 6,000 light years. It also possesses a spherical halo of stars, with a diameter of almost 100,000 light years. The total mass is usually estimated as about 100 thousand million solar masses, but some astronomers think there may be a good deal more mass in an extended halo. The solar system is some 30,000 light years from the center of the disk, and slightly "north" of the central plane of the disk. The disk rotates, with speeds ranging up to about 250 kilometers per second, and exhibits giant spiral arms. Altogether a glorious sight, if only we could see it from outside! The whole system is usually now called the Galaxy, or, taking a larger view, "our galaxy."

The other of the cosmologically interesting features of the night sky is much less obvious than the Milky Way. In the constellation Andromeda there is a hazy patch, not easy to see but clearly visible on a good night if you know where to look for it. The first written mention of this object appears to be a listing in the *Book of the Fixed Stars,* compiled in A.D. 964 by the Persian astronomer Abdurrahman Al-Sufi. He described it as a "little cloud." After telescopes became available, more and more such extended objects were discovered, and astronomers in the seventeenth and eighteenth centuries found that these objects were getting in the way of the search for things that seemed really interesting, the comets. In order to provide a convenient list of objects *not* to look at while hunting for comets, Charles Messier in 1781 published a celebrated catalog, *Nebulae and Star Clusters.* Astronomers still refer to the 103 objects in this catalog by their Messier numbers—thus the Andromeda Nebula is M31, the Crab Nebula is M1, and so on.

Even in Messier's time it was clear that these extended objects are not all the same. Some are obviously clusters of stars, like the Pleiades (M45). Others are irregular clouds of glowing gas, often colored, and often associated with one or more stars, like the Giant Nebula in Orion (M42). Today we know that objects of these two types are within our galaxy, and they need not concern us further here. However, about a third of the objects in Messier's catalog were white nebulae of a fairly regular elliptical shape, of which the most prominent was the Andromeda Nebula (M31). As telescopes improved, thousands more of these were found, and by the end of the nineteenth century spiral arms had been identified in some, including M31 and M33. However, the best telescopes of the eighteenth and nineteenth centuries were unable to resolve the elliptical or spiral nebulae into stars, and their nature remained in doubt.

It seems to have been Immanuel Kant who first proposed that some of the nebulae are galaxies like our own. Picking up Wright's theory of the Milky Way, Kant in 1755 in his *Universal Natural History and Theory of the Heavens* suggested that the nebulae "or rather a species of them" are really circular disks about the same size and shape as our own galaxy. They appear elliptical because most of them are viewed at a slant, and of course they are faint because they are so far away.

The idea of a universe filled with galaxies like our own became widely though by no means universally accepted by the beginning of the nineteenth century. However, it remained an open possibility that these elliptical and spiral nebulae might prove to be mere clouds within our own galaxy, like other objects in Messier's catalog. One great source of confusion was the observation of exploding stars in some of the spiral nebulae. If these nebulae were really independent galaxies, too far away for us to pick out individual stars, then the explosions would have to be incredibly powerful to be so bright at such a great distance. In this connection, I cannot resist quoting one example of nineteenth-century scientific prose at its ripest. Writing in 1893, the English historian of astronomy Agnes Mary Clerke remarked:

The well known nebula in Andromeda, and the great spiral in Canes Venatici are among the more remarkable of those giving a continuous spectrum; and as a general rule, the emissions of all such nebulae as present the appearance of star-clusters grown misty through excessive distance, are of the same kind. It would, however, be eminently rash to conclude thence that they are really aggregations of such sun-like bodies. The improbability of such an inference has been greatly enhanced by the occurrence, at an interval of a quarter of a century, of stellar outbursts in two of them. For it is practically certain that, however distant the nebulae, the stars were equally remote; hence, if the constituent particles of the former be suns, the incomparably vaster orbs by which their feeble light was well-nigh obliterated must, as was argued by Mr. Proctor, have been on a scale of magnitude such as the imagination recoils from contemplating.

Today we know that these stellar outbursts were indeed "on a scale of magnitude such as the imagination recoils from contemplating." They were supernovas, explosions in which one star approaches the luminosity of a whole galaxy. But this was not known in 1893.

The question of the nature of the spiral and elliptical nebulae could not be settled without some reliable method of determining how far away they are. Such a yardstick was at last discovered after the completion of the 100″ telescope at Mount Wilson, near Los Angeles. In 1923 Edwin Hubble was for the first time able to resolve the Andromeda Nebula into separate stars. He found that its spiral arms included a few bright variable stars, with the same sort of periodic variation of luminosity as was already familiar for a class of stars in our galaxy known as Cepheid variables. The reason this was so important was that in the preceding decade the work of Henrietta Swan Leavitt and Harlow Shapley of the Harvard College Observatory had provided a tight relation between the observed periods of variation of the Cepheids and their absolute luminosities. (Absolute luminosity is the total radiant power emitted by an astronomical object in all directions. Apparent luminosity is the radiant power received by us in each square centimeter of our telescope mirror. It is the apparent rather than the absolute luminosity that determines the subjective degree of brightness of astronomical objects. Of course, the apparent luminosity depends not only on the absolute luminosity, but also on the distance; thus, knowing both the absolute and the apparent luminosities of an astronomical body, we can infer its distance.) Hubble, observing the apparent luminosity of the Cepheids in the Andromeda Nebula, and estimating their absolute luminosity from their periods, could immediately calculate their distance, and hence the distance of the Andromeda Nebula, using the simple rule that apparent luminosity is proportional to the absolute luminosity and inversely proportional to the square of the distance. His conclusion was that the Andromeda Nebula is at a distance of 900,000 light years, or more than ten times farther than the most distant known objects in our own galaxy. Several recalibrations of the Cepheid period-luminosity relation by Walter Baade and others have by now increased the distance of the Andromeda Nebula to over two million light years, but the conclusion was already clear in 1923: the Andromeda Nebula, and the thousands of similar nebula, are galaxies like our own, filling the universe to great distances in all directions.

Even before the extragalactic nature of the nebulae had been settled, astronomers had been able to identify lines in their spectrum with known lines in familiar atomic spectra. However, it was discovered in the decade 1910–1920 by Vesto Melvin Slipher of the Lowell observatory that the spectral lines of many nebulae are shifted slightly to the red or blue. These shifts were immediately interpreted as due to a Doppler effect, indicating that the nebulae are moving away from or toward the earth. For instance, the Andromeda Nebula was found to be moving toward the earth at about 300 kilometers

per motion of Barnard's Star: The position of Barnard's star (indicated by white arrow) shown in two photographs taken 22 years apart. The change in the position of Barnard's star tive to the brighter background stars is readily apparent. In these 22 years, the direction Barnard's star changed by 3.7 minutes of arc; thus the "proper motion" is 0.17 minutes of per year. (Yerkes Observatory Photograph)

The Milky Way in Sagittarius: This photo shows the Milky Way in the direction of the center of our galaxy, in the constellation Sagittarius. The flatness of the galaxy is evident. The dark regions running through the plane of the Milky Way arise from clouds of dust, which absorb the light from the stars behind them. (Hale Observatories Photograph)

The Spiral Galaxy M104: This is a giant system of about a hundred billion stars, much like our own galaxy, but some 60 million light years away from us. From our viewpoint, M104 appears almost edge on, showing clearly the presence of both a bright spherical halo and a flat disc. The disc is marked with dark lanes of dust, much like the dusty regions of our own galaxy, as shown in the preceding photograph. This photograph was taken with the 60-inch reflector at Mount Wilson, California. (Yerkes Observatory Photograph)

The Great Galaxy M31 in Andromeda: This is the nearest large galaxy to our own. The tw
bright spots to the upper right and below the center are smaller galaxies, NGC 205 and 22
held in orbit by the gravitational field of M31. Other bright spots in the picture are foregrou
objects, stars within our own galaxy that happen to lie between the earth and M31. Th
picture was taken with the 48-inch telescope at Palomar. (Hale Observatories Photograph)

Detail of the Andromeda Galaxy: This shows one part of the Andromeda galaxy M31, corresponding to the lower-right-hand corner ("south–preceding region") in the preceding photograph. Taken with the 100-inch telescope at Mt. Wilson, this photograph has sufficient resolution to reveal separate stars in the spiral arms of M31. It was the study of such stars by Hubble in 1923 that showed conclusively that M31 is a galaxy more or less like our own, and not an outlying part of our galaxy. (Hale Observatories Photograph)

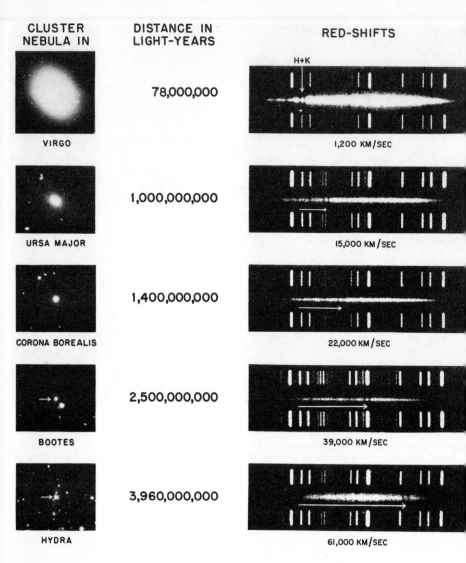

CLUSTER NEBULA IN	DISTANCE IN LIGHT-YEARS	RED-SHIFTS
VIRGO	78,000,000	1,200 KM/SEC
URSA MAJOR	1,000,000,000	15,000 KM/SEC
CORONA BOREALIS	1,400,000,000	22,000 KM/SEC
BOOTES	2,500,000,000	39,000 KM/SEC
HYDRA	3,960,000,000	61,000 KM/SEC

Relation Between Red Shift and Distance: Shown here are bright galaxies in five galaxy clusters together with their spectra. The spectra of the galaxies are the long, horizontal white smears crossed with a few short, dark vertical lines. Each position along these spectra corresponds to light from the galaxy with a definite wavelength; the dark vertical lines arise from absorption of light within the atmospheres of stars in these galaxies. (The bright vertical lines above and below each galaxy's spectrum are merely standard comparison spectra, superimposed on the spectrum of the galaxy to aid in determining wavelengths.) The arrows below each spectrum indicate the shift of two specific absorption lines (the H and K lines of calcium) from their normal position, toward the right (red) end of the spectrum. If interpreted as a Doppler effect, the red shift of these absorption lines indicates a velocity ranging from 1,200 kilometers per second for the Virgo cluster galaxy to 61,000 kilometers per second for the Hydra cluster. With a red shift proportional to distance, this indicates that these galaxies are at successively greater distances. (The distances given here are computed with a Hubble constant of 15.3 kilometers per second per million light years.) This interpretation is confirmed by the fact that the galaxies appear progressively smaller and dimmer with increasing red shift. (Hale Observatories Photograph) per second, while the more distant cluster of galaxies in the constellation Virgo were found to be moving away from the earth at about 1,000 kilometers per second.

At first it was thought that these might be merely relative velocities, reflecting a motion of our own solar system toward some galaxies and away from others. However, this explanation became untenable as more and more of the larger spectral shifts were discovered, all toward the red end of the spectrum. It appeared that aside from a few close neighbors like the Andromeda Nebula, the other galaxies are generally rushing away from our own. Of course, this does not mean that our galaxy has any special central position. Rather, it appears that the universe is undergoing some sort of explosion in which every galaxy is rushing away from every other galaxy.

This interpretation became generally accepted after 1929, when Hubble announced that he had discovered that the red shifts of galaxies increase roughly in proportion to the distance from us. The importance of this observation is that it is just what we should predict according to the simplest possible picture of the flow of matter in an exploding universe.

We would expect intuitively that at any given time the universe ought to look the same to observers in all typical galaxies, and in whatever directions they look. (Here, and below, I will use the label "typical" to indicate galaxies that do not have any large peculiar motion of their own, but are simply carried along with the general cosmic flow of galaxies.) This hypothesis is so natural (at least since Copernicus) that it has been called *the* Cosmological Principle by the English astrophysicist Edward Arthur Milne.

As applied to the galaxies themselves, the Cosmological Principle requires that an observer in a typical galaxy should see all the other galaxies moving with the same pattern of velocities, whatever typical galaxy the observer happens to be riding in. It is a direct mathematical consequence of this principle that the relative speed of any two galaxies must be proportional to the distance between them, just as found by Hubble.

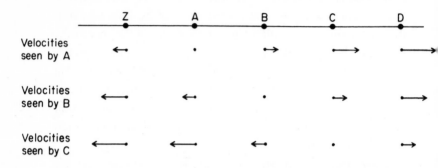

Figure 1. *Homogeneity and the Hubble Law.* A string of equally spaced galaxies Z, A, B, C, . . . are shown, with velocities as measured from A or B or C indicated by the lengths and directions of the attached arrows. The principle of homogeneity requires that the velocity of C as seen by B is equal to the velocity of B as seen by A; adding these two velocities gives the velocity of C as seen by A, indicated by an arrow twice as long. Proceeding in this way, we can fill out the whole pattern of velocities shown in the figure. As can be seen, the velocities obey the Hubble law: the velocity of any galaxy as seen by any other is proportional to the distance between them. This is the only pattern of velocities consistent with the principle of homogeneity.

To see this, consider three typical galaxies A, B, and C, strung out in a straight line. (See figure 1.) Suppose that the distance between A and B is the same as the distance between B and C. Whatever the speed of B as seen from A, the Cosmological Principle requires that C should have the same speed relative to B. But note then that C, which is twice as far away from A as is B, is also moving twice as fast relative to A as is B. We can add more galaxies in our chain, always with the result that the speed of recession of any galaxy relative to any other is proportional to the distance between them.

As often happens in science, this argument can be used both forward and backward. Hubble, in observing a proportionality between the distances of galaxies and their speeds of recession, was indirectly verifying the truth of the Cosmological Principle. This is enormously satisfying philosophically—why should any part of the universe or any direction be any different from any other? It also helps to reassure us that the astronomers really are looking at some appreciable part of the universe, not a mere local eddy in a vaster cosmic maelstrom. Contrariwise, we can take the Cosmological Principle for granted on *a priori* grounds, and deduce the relation of proportionality between distance and velocity, as done in the last paragraph. In this way, through the relatively easy measurement of Doppler shifts, we are able to judge the distance of very remote objects from their velocities.

The Cosmological Principle has observational support of another sort, apart from the measurement of Doppler shifts. After making due allowances for the distortions due to our own galaxy and the rich nearby cluster of galaxies in the constellation Virgo, the universe seems remarkably isotropic; that is, it looks the same in all directions. (This is shown even more convincingly by the microwave background radiation discussed in the next chapter.) But ever since Copernicus we have learned to beware of supposing that there is anything special about mankind's location in the universe. So if the universe is isotropic around us, it ought to be isotropic about every typical galaxy. However, any point in the universe can be carried into any other point by a series of rotations around fixed centers (see figure 2), so if the universe is isotropic around every point, it is necessarily also homogeneous.

Before going any further, a number of qualifications have to be attached to the Cosmological Principle. First, it is ob-

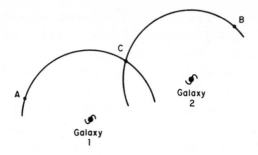

Figure 2. *Isotropy and Homogeneity*. If the universe is isotropic about both galaxy 1 and galaxy 2, then it is homogeneous. In order to show that conditions at two arbitrary points A and B are the same, draw a circle through A around galaxy 1, and another circle through B around galaxy 2. Isotropy around galaxy 1 requires that conditions are the same at A and at the point C where the circles intersect. Likewise, isotropy around galaxy 2 requires that conditions are the same at B and C. Hence they are the same at A and B.

viously not true on small scales—we are in a galaxy which belongs to a small local group of other galaxies (including M31 and M33), which in turn lies near the enormous cluster of galaxies in Virgo. In fact, of the 33 galaxies in Messier's catalog, almost half are in one small part of the sky, the constellation Virgo. The Cosmological Principle, if at all valid, comes into play only when we view the universe on a scale at least as large as the distance between clusters of galaxies, or about 100 million light years.

There is another qualification. In using the Cosmological Principle to derive the relation of proportionality between galactic velocities and distances, we supposed that if the velocity of C relative to B is the same as the velocity of B relative to A, then the velocity of C relative to A is twice as great. This is just the usual rule for adding velocities with which everyone is familiar, and it certainly works well for the relatively low velocities of ordinary life. However, this rule must break down for velocities approaching the speed of light (300,000 kilometers per second), for otherwise, by adding up a number of relative velocities, we could achieve a total velocity greater than that of light, which is forbidden by Einstein's Special Theory of Relativity. For instance, the usual rule for addition of velocities would say that if a passenger on an airplane moving at three-quarters the speed of light fires a bullet forward at three-quarters the speed of light, then the speed of the bullet relative to the ground is one and one-half times the speed of light, which is impossible. Special relativity avoids this problem by changing the rule for adding velocities: the velocity of C relative to A is actually somewhat *less* than the sum of the velocities of B relative to A and C relative to B, so that no matter how many times we add together velocities less than that of light, we never get a velocity greater than that of light.

None of this was a problem for Hubble in 1929; none of the galaxies he studied then had a speed anywhere near the speed of light. Nevertheless, when cosmologists think about the really large distances characteristic of the universe as a whole, they must work in a theoretical framework capable of dealing with velocities approaching that of light, that is, Einstein's Special and General Theories of Relativity. Indeed, when we deal with distances this great, the concept of distance itself becomes ambiguous, and we must specify whether we mean distance as a measured by observation of luminosities, or diameters, or proper motions, or something else.

Returning now to 1929: Hubble estimated the distance to 18 galaxies from the apparent luminosity of their brightest stars, and compared these distances with the galaxies' respective velocities, determined spectroscopically from their Doppler shifts. His conclusion was that there is a "roughly linear relation" (i.e., simple proportionality) between velocities and distances. Actually, a look at Hubble's data leaves me perplexed how he could reach such a conclusion—galactic velocities seem almost uncorrelated with their distance, with only a mild tendency for velocity to increase with distance. In fact, we would not *expect* any neat relation of proportionality between velocity and distance for these 18 galaxies—they are all much too close, none being farther than the Virgo cluster. It is difficult to avoid the conclusion that, relying either on the simple arguments sketched above or the related theoretical developments to be discussed below, Hubble knew the answer he wanted to get.

However that may be, by 1931 the evidence had greatly improved, and Hubble was able to verify the proportionality between velocity and distance for galaxies with velocities ranging up to 20,000 kilometers per second. With the estimates of distance then available, the conclusion was that velocities increase by 170 kilometers per second for every million light years distance; thus, a velocity of 20,000 kilometers per second means a distance of 120 million light years. This figure, of a certain velocity increase per distance, is generally known as the "Hubble constant." (It is a constant in the sense that the proportionality between velocity and distance is the same for all galaxies at a given time, but, as we shall see, the Hubble constant changes with time as the universe evolves.)

By 1936 Hubble, working with the spectroscopist Milton Humason, was able to measure the distance and velocity of the Ursa Major II cluster of galaxies. It was found to be receding at a speed of 42,000 kilometers per second—14 percent of the speed of light. The distance, then estimated as 260 million light years, was at the limit of Mt. Wilson's capability, and Hubble's work had to stop. With the advent after the war of larger telescopes at Palomar and Mt. Hamilton, Hubble's program was taken up again by other astronomers (notably Allan Sandage of Palomar and Mt. Wilson), and continues to the present time.

The conclusion generally drawn from this half-century of observation is that the galaxies are receding from us, with speeds proportional to the distance (at least for speeds not too close to that of light). Of course, as already emphasized in our discussion of the Cosmological Principle, this does not mean that we are in any specially favored or unfavored position in the cosmos; *every* pair of galaxies is moving apart at a relative speed proportional to their separation. The most important modification of Hubble's original conclusions is a revision of the extragalactic distance scale: partly as a result of a recalibration of the Leavitt-Shapley Cepheid period-luminosity relation by Walter Baade and others, the distances to far galaxies are now estimated to be about ten times larger than was thought in Hubble's time. Thus, the Hubble constant is now believed to be only about 15 kilometers per second per million light years.

What does all this say about the origin of the universe? If the galaxies are rushing apart, then they must once have been closer together. To be specific, if their velocity has been constant, then the time it has taken any pair of galaxies to reach their present separation is just the present distance between them divided by their relative velocity. But with a velocity which is proportional to their present separation, this time is the same for any pair of galaxies—they must have all been close together at the same time in the past! Taking the Hubble constant as 15 kilometers per second per million light years, the time since the galaxies began to move apart would be a million light years divided by 15 kilometers per second, or 20 thousand million years. We shall refer to the "age" calculated in this way as the "characteristic expansion time"; it is simply the reciprocal of the Hubble constant. The true age of the universe is actually *less* than the characteristic expansion time because, as we shall see, the galaxies have not been moving at constant velocities, but have been slowing down under the influence of their mutual gravitation. Therefore, if the Hubble constant is 15 kilometers per second per million light years, the age of the universe must be less than 20,000 million years.

Sometimes we summarize all this by saying briefly that the size of the universe is increasing. This does not mean that the universe necessarily has a finite size, although it well may have. This language is used because in any given time, the separation between any pair of typical galaxies increases by the same *fractional* amount. During any interval that is short enough so that the galaxies' velocities remain approximately constant, the increase in the separation between a pair of typical galaxies will be given by the product of their relative velocity and the elapsed time, or, using the Hubble law, by the product of the Hubble constant, the separation, and the time. But then the *ratio* of the increase in separation to the separation itself will be given by the Hubble constant times the elapsed time, which is the same for any pair of galaxies. For instance, during a time interval of 1 percent of the characteristic expansion time (the reciprocal of the Hubble constant), the separation of every pair of typical galaxies will increase by 1 percent. We would then, speaking loosely, say that the size of the universe has increased by 1 percent.

I do not want to give the impression that everyone agrees with this interpretation of the red shift. We do not actually observe galaxies rushing away from us; all we are sure of is that the lines in their spectra are shifted to the red, i.e., toward longer wavelengths. There are eminent astronomers who doubt that the red shifts have anything to do with Doppler shifts or with an expansion of the universe. Halton Arp, of the Hale Observatories, has emphasized the existence of groupings of galaxies in the sky in which some galaxies have very different red shift from the others; if these groupings represent true physical associations of neighboring galaxies, they could hardly have grossly different velocities. Also, it was discovered by Maarten Schmidt in 1963 that a certain class of objects which have the appearance of stars nevertheless have enormous red shifts, in some cases over 300 percent! If these "quasi-stellar objects" are as far away as their red shifts indicate, they must be emitting enormous amounts of energy to be so bright. Finally, it is not easy to determine the relation between velocity and distance at really large distances.

There is, however, an independent way to confirm that the galaxies are really moving apart, as indicated by the red shifts. As we have seen, this interpretation of the red shifts implies that the expansion of the universe began somewhat less than 20 thousand million years ago. It will therefore tend to be confirmed if we can find any other evidence that the universe is actually that old. In fact, there is a good deal of evidence that our galaxy is about 10–15 thousand million years old. This estimate comes both from the relative abundance of various radioactive isotopes in the earth (especially the uranium isotopes, U-235 and U-238) and from calculation of the evolution of stars. There is certainly no direct connection between the rates of radioactivity or stellar evolution and the red shift of distant galaxies, so the presumption is strong that the age of the universe deduced from the Hubble constant really does represent a true beginning.

In this connection, it is historically interesting to recall that during the 1930s and 1940s the Hubble constant was believed to be much larger, about 170 kilometers per second per million light years. By our previous reasoning the age of the universe would then have to be one million light years divided by 170 kilometers per second, which is about 2,000 million years, or even less if we take gravitational braking into account. But it has been well known since the studies of radioactivity by Lord Rutherford that the earth is much older than this; it is now thought to be about 4,600 million years old! The earth can hardly be older than the universe, so astronomers were forced to doubt that the red shift really tells us anything about the age of the universe. Some of the most ingenious cosmological ideas of the 1930s and 1940s were generated by this apparent paradox, including perhaps the steady-state theory. It may be that the removal of the age paradox by the tenfold expansion of the extragalactic distance scale in the 1950s was the essential precondition for the emergence of the big bang cosmology as a standard theory.

The picture of the universe we have been developing here is of an expanding swarm of galaxies. Light has so far played for us only the role of a "starry messenger," carrying information of the galaxies' distance and velocity. However, conditions were very different in the early universe; as we shall see, it was light that then formed the dominant constituent of the universe, and ordinary matter played only the role of a negligible contamination. It will therefore be useful to us later if we restate what we have learned about the red shift in terms of the behavior of light waves in an expanding universe.

Consider a light wave traveling between two typical galaxies. The separation between the galaxies equals the light travel time times the speed of light, while the increase in this separation during the light's journey equals the light travel time times the galaxies' relative velocity. When we calculate the *fractional* increase in separation, we divide the increase in separation by the mean value of this separation during the increase, and we find that the light travel time cancels out: the fractional increase in separation of these two galaxies (and hence of any other typical galaxies) during the light travel time is just the ratio of the galaxies' relative velocity to the speed of light. But as we have seen earlier, this same ratio also gives the fractional increase in the wavelength of the light wave during its journey. Thus, *the wavelength of any ray of light simply increases in proportion to the separation between typical galaxies as the universe expands.* We can think of the wave crests as being "pulled" farther and farther apart by the expansion of the universe. Although our argument has been strictly valid only for short travel times, by putting together a sequence of these trips we can conclude that the same is true in general. For instance, when we look at the galaxy 3C295, and find that the wavelengths in its spectra are 46 percent larger than in our standard tables of spectral wavelengths, we can conclude that the universe is now 46 percent larger than it was when the light left 3C295.

Up to this point, we have concerned ourselves with matters that physicists call "kinematic," having to do with the description of motion apart from any consideration of the forces that govern it. However, for centuries physicists and astronomers have also tried to understand the dynamics of the universe. Inevitably this has led to a study of the cosmological role of the only force that acts between astronomical bodies, the force of gravitation.

As might be expected, it was Isaac Newton who first grappled with this problem. In a famous correspondence with the Cambridge classicist Richard Bentley, Newton admitted that if the matter of the universe were evenly distributed in a *finite* region, then it would all tend to fall toward the center, "and there compose one great spherical mass." On the other hand, if matter were evenly dispersed through an *infinite* space, there would be no center to which it could fall. It might in this case contract into an infinite number of clumps, scattered through the universe; Newton suggested that this might even be the origin of the sun and stars.

The difficulty of dealing with the dynamics of an infinite medium pretty well paralyzed further progress until the advent of general relativity. This is no place to explain general relativity, and in any case it turned out to be less important to cosmology than was at first thought. Suffice it to say that Albert Einstein used the existing mathematical theory of non-Euclidean geometry to account for gravitation as an effect of the curvature of space and time. In 1917, a year after the completion of his general theory of relativity, Einstein tried to find a solution of his equations that would describe the space-time geometry of the whole universe. Following the cosmological ideas then current, Einstein looked specifically for a solution that would be homogeneous, isotropic, and, unfortunately, *static*. However, no such solution could be found. In order to achieve a model that fit these cosmological presuppositions, Einstein was forced to mutilate his equations by introducing a term, the so-called cosmological constant, which greatly marred the elegance of the original theory, but which could serve to balance the attractive force of gravitation at large distances.

Einstein's model universe was truly static, and predicted no red shifts. In the same year, 1917, another solution of Einstein's modified theory was found by the Dutch astronomer W. de Sitter. Although this solution appeared to be static, and was therefore acceptable according to the cosmological ideas of the times, it had the remarkable property of predicting a red shift proportional to the distance! The existence of large nebular red shifts was not then known to European astronomers. However, at the end of World War I news of the observation of large red shifts reached Europe from America, and de Sitter's model acquired instant celebrity. In fact, in 1922 when the English astronomer Arthur Eddington wrote the first comprehensive treatise on general relativity, he analyzed the existing red-shift data in terms of the de Sitter model. Hubble himself said that it was the de Sitter model that drew astronomers' attention to the importance of a dependence of red shift on distance, and this model may have been in the back of his mind when he discovered the proportionality of red shift to distance in 1929.

Today this emphasis on the de Sitter model seems misplaced. For one thing, it is not really a static model at all—it looked static because of the peculiar way that spatial coordinates were introduced, but the distance between "typical" observers in the model actually increases with time, and it is this general recession that produces the red shift. Also, the reason that the red shift turned out to be proportional to the distance in de Sitter's model is just that this model satisfies the Cosmological Principle, and, as we have seen, we expect a proportionality between relative velocity and distance in *any* theory that satisfies this principle.

At any rate, the discovery of the recession of distant galaxies soon aroused interest in cosmological models that are homogeneous and isotropic but not static. A "cosmological constant" was then not needed in the field equations of gravitation, and Einstein came to regret that he had ever considered any such change in his original equations. In 1922 the general homogeneous and isotropic solution of the original Einstein equations was found by the Russian mathematician Alexandre Friedmann. It is these Friedmann models, based on the original Einstein field equations, and not the Einstein or de Sitter models, that provide the mathematical background for most modern cosmological theories.

The Friedmann models are of two very different types. If the average density of the matter of the universe is *less* than or equal to a certain critical value, then the universe must be spatially infinite. In this case the present expansion of the universe will go on forever. On the other hand, if the density of the universe is *greater* than this critical value, then the gravitational field produced by the matter curves the universe back on itself; it is finite though unbounded, like the surface of a sphere. (That is, if we set off on a journey in a straight line, we do not reach any sort of edge of the universe, but simply come back to where we began.) In this case the gravitational fields are strong enough eventually to stop the expansion of the universe, so that it will eventually implode back to indefinitely large density. The critical density is proportional to the square of the Hubble constant; for the presently popular value of 15 kilometers per second per million light years, the critical density equals 5×10^{-30} grams per cubic centimeter, or about three hydrogen atoms per thousand liters of space.

The motion of any typical galaxy in the Friedmann models is precisely like that of a stone thrown upward from the surface of the earth. If the stone is thrown fast enough or, what amounts to the same thing, if the mass of the earth is small enough, then the stone will gradually slow down, but will nevertheless escape to infinity. This corresponds to the case of a cosmic density less than the critical density. On the other hand, if the stone is thrown with insufficient speed, then it will rise to a maximum height and then plunge back downward. This of course corresponds to a cosmic density above the critical density.

This analogy makes clear why it was not possible to find static cosmological solutions of Einstein's equations—we might not be too surprised to see a stone rising from or falling to the surface of the earth, but we would hardly expect to find one hanging still in midair. The analogy also helps us to avoid a common misconception about the expanding universe. The galaxies are not rushing apart because of some mysterious force that is pushing them apart, just as the rising stone in our analogy is not being repelled by the earth. Rather, the galaxies are moving apart because they were thrown apart by some sort of explosion in the past.

It was not realized in the 1920s, but many of the detailed properties of the Friedmann models can be calculated quantitatively using this analogy, without any reference to general relativity. In order to calculate the motion of any typical galaxy relative to our own, draw a sphere with us at the center and the galaxy of interest on the surface; the motion of this galaxy is precisely the same as if the mass of the universe consisted only of the matter within this sphere, with nothing outside. It is just as if we dug a cave deep in the interior of the earth, and observed the way that bodies fall—we would find that the gravitational acceleration toward the center depended only on the amount of matter closer to the center than our cave, as if the surface of the earth were just at the depth of our cave. This remarkable result is embodied in a theorem, valid in both Newton's and Einstein's theories of gravitation, which depends only on the spherical symmetry of the system under study; the general relativistic version of this theorem was proved by the American mathematician G. D. Birkhoff in 1923, but its cosmological significance was not realized for some decades after.

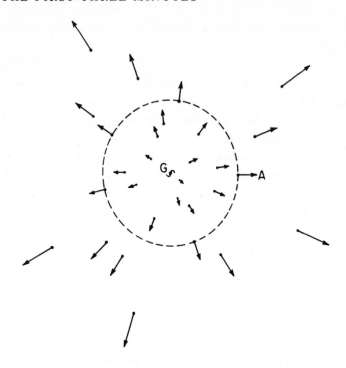

Figure 3 *Birkhoff's Theorem and the Expansion of the Universe.* A number of galaxies are shown, together with their velocities relative to a given galaxy G, indicated here by the lengths and directions of the attached arrows. (In accordance with the Hubble law, these velocities are taken to be proportional to the distance from G.) Birkhoff's theorem states that in order to calculate the motion of a galaxy A relative to G, it is only necessary to take into account the mass contained within the sphere around G that passes through A, shown here by the dashed line. If A is not too far from G, the gravitational field of the matter within the sphere will be moderate, and the motion of A can be calculated by the rules of Newtonian mechanics.

We can employ this theorem to calculate the critical density of the Friedmann models. (See figure 3.) When we draw a sphere with us at the center and some distant galaxy on the surface, we can use the mass of the galaxies within the sphere to calculate an escape velocity, the velocity which a galaxy at the surface would have to have to be able just barely to escape to infinity. It turns out that this escape velocity is proportional to the radius of the sphere—the more massive the sphere, the faster one must go to escape it. But the Hubble law tells us that the actual velocity of a galaxy on the surface of the sphere is also proportional to the radius of the sphere—the distance from us. Thus, although the escape velocity depends on the radius, the *ratio* of the galaxy's actual velocity to its escape velocity does not depend on the size of the sphere; it is the same for all galaxies, and it is the same whatever galaxy we take as the center of the sphere. Depending on the values of the Hubble constant and the cosmic density, *every* galaxy which moves according to the Hubble law will either exceed escape velocity and escape to infinity, or will fall short of escape velocity and fall back toward us at some time in the future. The critical density is simply the value of the cosmic density at which the escape velocity of each galaxy just equals the velocity given by Hubble's law. The critical density can only depend on the Hubble constant, and, in fact, it turns out to be simply proportional to the square of the Hubble constant. (See mathematical note, p. 2, p. 167.)

The detailed time dependence of the size of the universe (that is, the distance between any typical galaxies) can be worked out using similar arguments, but the results are rather complicated. (See figure 4.) However, there is one simple result that will be very important to us later on. In the early era of the universe, the size of the universe varied as a simple power of time: the two-thirds power if the density of radiation could be neglected, or the one-half power if the density of radiation exceeded that of matter. (See mathematical note 3, p. 171). The one aspect of the Friedmann cosmological models that cannot be understood without general relativity is the relation between the density and the geometry—the universe

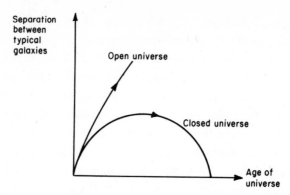

Figure 4. *Expansion and Contraction of the Universe.* The separation between typical galaxies is shown (in arbitrary units) as a function of time, for two possible cosmological models. In the case of an "open universe," the universe is infinite; the density is less than the critical velocity; and the expansion, though slowing down, will continue forever. In the case of a "closed universe," the universe is finite; the density is greater than the critical density; and the expansion will eventually cease and be followed by a contraction. These curves are calculated using Einstein's field equations without a cosmological constant, for a matter-dominated universe.

is open and infinite or closed and finite according to whether the velocity of galaxies is greater or less than the escape velocity.

One way to tell whether or not the galactic velocities exceed escape velocity is to measure the rate at which they are slowing down. If this deceleration is less (or greater) than a certain amount, then escape velocity is (or is not) exceeded. In practice, this means that one must measure the curvature of the graph of red shift versus distance for very distant galaxies. (See figure 5.) As one proceeds from a more dense finite universe to a less dense infinite universe, the curve of red shift versus distance flattens out at very large distances. The study of the shape of the red shift-distance curve at great distances is often called the "Hubble program."

A tremendous effort has been put into this program by Hubble, Sandage, and recently others as well. So far the re-

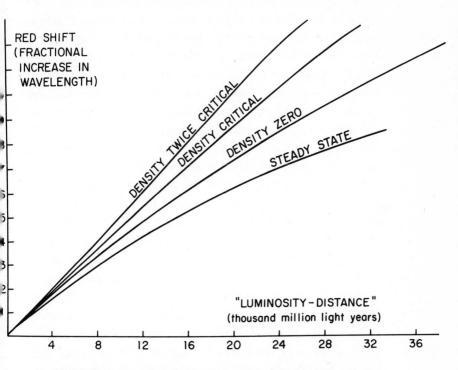

Figure 5. *Red Shift vs. Distance.* The red shift is shown here as a function of distance, for four possible cosmological theories. (To be precise, the "distance" here is "luminosity distance"—the distance inferred for an object of known intrinsic or absolute luminosity from observations of its apparent luminosity.) The curves labeled "density twice critical," "density critical," and "density zero" are calculated in the Friedmann model, using Einstein's field equations for a matter-dominated universe, without a cosmological constant; they correspond respectively to a universe that is closed, just barely open, or open. (See figure 4.) The curve marked "steady state" will apply to any theory in which the appearance of the universe does not change with time. Current observations are not in good agreement with the "steady-state" curve, but they do not definitely decide among the other possibilities, because in non-steady-state theories galactic evolution makes determination of distance very problematical. All curves are drawn with the Hubble constant taken as 15 kilometers per second per million light years (corresponding to a characteristic expansion time of 20,000 million years), but the curves can be used for any other value of the Hubble constant by simply rescaling all distances.

sults have been quite inconclusive. The trouble is that in estimating the distance to far galaxies it is impossible to pick out Cepheid variables or brightest stars to use as distance indicators; rather, we must estimate the distance from the apparent luminosity of the galaxies themselves. But how do we know that the galaxies we study all have the same *absolute* luminosity? (Recall that apparent luminosity is the radiant power received by us per unit telescope area, while absolute luminosity is the total power emitted in all directions by the astronomical object; apparent luminosity is proportional to absolute luminosity and inversely proportional to the square of the distance.) There are terrible dangers from selection effects—as we look out farther and farther, we tend to pick out galaxies of greater and greater absolute luminosity. An even worse problem is galactic evolution. When we look at very distant galaxies we see them as they were thousands of millions of years ago, when the light rays started on their journey to us. If typical galaxies were brighter then than now, we will underestimate their true distance. One possibility, raised very recently by J. P. Ostriker and S. D. Tremaine of Princeton, is that the larger galaxies evolve not only because their individual stars evolve, but also because they gobble up small neighboring galaxies! It is going to be a long time before we can be sure that we have an adequate quantitative understanding of these various kinds of galactic evolution.

At present, the best inference that can be drawn from the Hubble program is that the deceleration of distant galaxies seems fairly small. This would mean that they are moving at more than escape velocity, so that the universe is open and will go on expanding forever. This fits in well with estimates of the cosmic density; the visible matter in galaxies seems to add up to not more than a few percent of the critical density. However, about this too there is uncertainty. Estimates of galactic mass have been increasing in recent years. Also, as suggested by George Field of Harvard and by others, there may be an intergalactic gas of ionized hydrogen which could provide a critical cosmic density of matter and yet have escaped detection.

Fortunately, it is not necessary to come to a definite decision about the large-scale geometry of the universe in order to draw conclusions about its beginning. The reason is that the universe has a sort of horizon, and this horizon shrinks rapidly as we look back toward the beginning.

No signal can travel faster than the speed of light, so at any time we can only be affected by events occurring close enough so that a ray of light would have had time to reach us since the beginning of the universe. Any event that occurred beyond this distance could as yet have no effect on us—it is beyond the horizon. If the universe is now 10,000 million years old, the horizon is now at a distance of 30,000 million light years. But when the universe was a few minutes old, the horizon was at a distance of only a few light minutes—less than the present distance from the earth to the sun. It is true also that the whole universe was smaller then, in our agreed sense that the separation between any pair of bodies was less than now. However, as we look back toward the beginning, the distance to the horizon shrinks faster than the size of the universe. The size of the universe is proportional to the one-half or two-thirds power of the time (see mathematical note 3, p. 171), while the distance to the horizon is simply proportional to the time, so for earlier and earlier times, the horizon encloses a smaller and smaller portion of the universe. (See figure 6.)

As a consequence of this closing in of horizons in the early universe, the curvature of the universe as a whole makes less and less difference as we look back to earlier and earlier times.

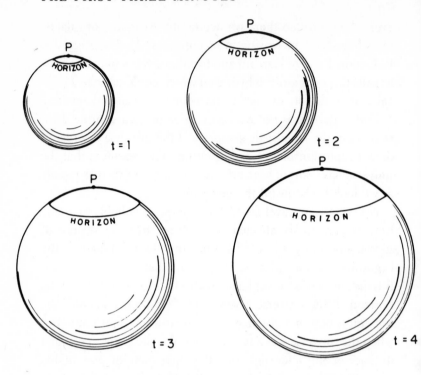

Figure 6. *Horizons in an Expanding Universe.* The universe is symbolized here as a sphere, at four moments separated by equal time intervals. The "horizon" of a given point P is the distance from beyond which light signals would not have had time to reach P. The part of the universe within the horizon is indicated here by the unshaded cap on the sphere. The distance from P to the horizon grows in direct proportion to the time. On the other hand, the "radius" of the universe grows like the square root of the time, corresponding to the case of a radiation-dominated universe. In consequence, at earlier and earlier times, the horizon encloses a smaller and smaller proportion of the universe.

Thus, even though present cosmological theory and astronomical observation have not yet revealed the extent or the future of the universe, they give a pretty clear picture of its past.

The observations discussed in this chapter have opened to us a view of the universe that is as simple as it is grand. The universe is expanding uniformly and isotropically—the same pattern of flow is seen by observers in all typical galaxies, and in all directions. As the universe expands, the wavelengths of light rays are stretched out in proportion to the distance between the galaxies. The expansion is not believed to be due to any sort of cosmic repulsion, but is rather just the effect of the velocities left over from a past explosion. These velocities are gradually slowing down under the influence of gravitation; this deceleration appears to be quite slow, suggesting that the matter density of the universe is low and its gravitational field is too weak either to make the universe spatially finite or eventually to reverse the expansion. Our calculations allow us to extrapolate the expansion of the universe backward in time, and reveal that the expansion must have begun between 10,000 and 20,000 million years ago.

III

THE COSMIC MICROWAVE RADIATION BACKGROUND

T HE STORY told in the last chapter is one with which the astronomers of the past would have felt at home. Even the setting is familiar: great telescopes exploring the night sky from mountain tops in California or Peru, or the naked-eye observer in his tower, to "oft out-watch the Bear." As I mentioned in the Preface, this is also a story that has been told many times before, often in greater detail than here.

Now we come to a different kind of astronomy, to a story that could not have been told a decade ago. We will be dealing not with observations of light emitted in the last few hundred million years from galaxies more or less like our own, but with observations of a diffuse background of radio static left over from near the beginning of the universe. The setting also changes, to the roofs of university physics buildings, to balloons or rockets flying above the earth's atmosphere, and to the fields of northern New Jersey.

In 1964 the Bell Telephone Laboratory was in possession of an unusual radio antenna on Crawford Hill at Holmdel, New Jersey. The antenna had been built for communication via the *Echo* satellite, but its characteristics—a 20-foot horn reflector with ultralow noise—made it a promising instrument for radio astronomy. A pair of radio astronomers, Arno A. Penzias and Robert W. Wilson, set out to use the antenna to measure the intensity of the radio waves emitted from our galaxy at high galactic latitudes, i.e., out of the plane of the Milky Way.

This kind of measurement is very difficult. The radio waves from our galaxy, as from most astronomical sources, are best described as a sort of *noise*, much like the "static" one hears on a radio set during a thunderstorm. This radio noise is not easily distinguished from the inevitable electrical noise that is produced by the random motions of electrons within the radio antenna structure and the amplifier circuits, or from the radio noise picked up by the antenna from the earth's atmosphere. The problem is not so serious when one is studying a relatively "small" source of radio noise, like a star or a distant galaxy. In this case one can switch the antenna beam back and forth between the source and the neighboring empty sky; any spurious noise coming from the antenna structure, amplifier circuits, or the earth's atmosphere will be about the same whether the antenna is pointed at the source or the nearby sky, so it would cancel out when the two are compared. However, Penzias and Wilson were intending to measure the radio noise coming from our own galaxy—in effect, from the sky itself. It was therefore crucially important to identify any electrical noise that might be produced within their receiving system.

Previous tests of this system had in fact revealed a little more noise than could be accounted for, but it seemed likely that this discrepancy was due to a slight excess of electrical noise in the amplifier circuits. In order to eliminate such problems, Penzias and Wilson made use of a device known as a "cold load"—the power coming from the antenna was compared with the power produced by an artificial source cooled with liquid helium, about four degrees above absolute zero. The electrical noise in the amplifier circuits would be the same in both cases, and would therefore cancel out in the comparison, allowing a direct measurement of the power coming from the antenna. The antenna power measured in this way would consist only of contributions from the antenna structure, from the earth's atmosphere, and from any astronomical sources of radio waves.

Penzias and Wilson expected that very little electrical noise would be produced within the antenna structure. However, in order to check this assumption, they started their observations at a relatively short wavelength of 7.35 centimeters, where the radio noise from our galaxy should have been negligible. Some radio noise could naturally be expected at this wavelength from our earth's atmosphere, but this would have a characteristic dependence on direction: it would be proportional to the thickness of atmosphere along the direction in which the antenna was pointed—less toward the zenith, more toward the horizon. It was expected that, after subtraction of an atmospheric term with this characteristic dependence on direction, there would be essentially no antenna power left over, and this would confirm that the electrical noise produced within the antenna structure was indeed negligible. They would then be able to go on to study the galaxy itself at a longer wavelength, around 21 centimeters, where the galactic radio noise was expected to be appreciable.

(Incidentally, radio waves with wavelengths like 7.35 centimeters or 21 centimeters, and up to 1 meter, are known as "microwave radiation." This is because these wavelengths are shorter than those of the VHF band used by radar at the beginning of World War II.)

To their surprise, Penzias and Wilson found in the spring of 1964 that they were receiving a sizable amount of microwave noise at 7.35 centimeters that was independent of direction. They also found that this "static" did not vary with the time of day or, as the year went on, with the season. It did not seem that it could be coming from our galaxy; if it were, then the great galaxy M31 in Andromeda, which is in most respects similar to our own, would presumably also be radiating strongly at 7.35 centimeters, and this microwave noise would already have been observed. Above all, the lack of any variation of the observed microwave noise with direction indicated very strongly that these radio waves, if real, were not coming from the Milky Way, but from a much larger volume of the universe.

Clearly, it was necessary to reconsider whether the antenna itself might be producing more electrical noise than expected. In particular, it was known that a pair of pigeons had been roosting in the antenna throat. The pigeons were caught; mailed to the Bell Laboratories Whippany site; released; found back in the antenna at Holmdel a few days later; caught again; and finally discouraged by more decisive means. However, in the course of their tenancy, the pigeons had coated the antenna throat with what Penzias delicately calls "a white dielectric material," and this material might at room temperature be a source of electrical noise. In early 1965 it became possible to dismantle the antenna throat and clean out the mess, but this, and all other efforts, produced only a very small decrease in the observed noise level. The mystery remained: Where was this microwave noise coming from?

The one piece of numerical data that was available to Penzias and Wilson was the intensity of the radio noise they had observed. In describing this intensity they used a language that is common among radio engineers, but which turned out in this case to have unexpected relevance. Any sort of body at any temperature above absolute zero will always emit radio noise, produced by the thermal motions of electrons within the body. Inside a box with opaque walls, the intensity of the radio noise at any given wavelength depends only on the temperature of the walls—the higher the temperature, the more intense the static. Thus, it is possible to describe the intensity of radio noise observed at a given wavelength in terms of an "equivalent temperature"—the temperature of the walls of a box within which the radio noise would have the observed intensity. Of course, a radio telescope is not a thermometer; it measures the strength of radio waves by recording the tiny electric currents that the waves induce in the structure of the antenna. When a radio astronomer says that he observes radio noise with such and such an equivalent temperature, he means only that this is the temperature of the opaque box into which the antenna would have to be placed to produce the observed radio noise intensity. Whether or not the antenna *is* in such a box is of course another question.

(To forestall objections from experts, I should mention that radio engineers often describe the intensity of radio noise in terms of a so-called antenna temperature, which is slightly different from the "equivalent temperature" described above. For the wavelengths and intensities observed by Penzias and Wilson, the two definitions are virtually identical.)

Penzias and Wilson found that the equivalent temperature

of the radio noise they were receiving was about 3.5 degrees Centigrade above absolute zero (or more accurately, between 2.5 and 4.5 degrees above absolute zero). Temperatures measured on the Centigrade scale, but referred to absolute zero rather than the melting point of ice, are reported in "degrees Kelvin." Thus, the radio noise observed by Penzias and Wilson could be described as having an "equivalent temperature" of 3.5 degrees Kelvin, or 3.5° K for short. This was much greater than expected, but still very low in absolute terms, so it is not surprising that Penzias and Wilson brooded over their result for a while before publishing it. It certainly was not immediately clear that this was the most important cosmological advance since the discovery of the red shifts.

The meaning of the mysterious microwave noise soon began to be clarified through the operation of the "invisible college" of astrophysicists. Penzias happened to telephone a fellow radio astronomer, Bernard Burke of M.I.T., about other matters. Burke had just heard from yet another colleague, Ken Turner of the Carnegie Institution, of a talk that Turner had in turn heard at Johns Hopkins, given by a young theorist from Princeton, P. J. E. Peebles. In this talk Peebles argued that there ought to be a background of radio noise left over from the early universe, with a present equivalent temperature of roughly 10° K. Burke already knew that Penzias was measuring radio noise temperatures with the Bell Laboratories horn antenna, so he took the occasion of the telephone conversation to ask how the measurements were going. Penzias said that the measurements were going fine, but that there was something about the results he didn't understand. Burke suggested to Penzias that the physicists at Princeton might have some interesting ideas on what it was that his antenna was receiving.

c

In his talk, and in a preprint written in March 1965, Peebles had considered the radiation that might have been present in the early universe. "Radiation" is of course a general term, encompassing electromagnetic waves of all wavelengths—not only radio waves, but infrared light, visible light, ultraviolet light, X rays, and the very short-wavelength radiation called gamma rays. (See table, p. 156.) There are no sharp distinctions; with changing wavelength one kind of radiation blends gradually into another. Peebles noted that if there had not been an intense background of radiation present during the first few minutes of the universe, nuclear reactions would have proceeded so rapidly that a large fraction of the hydrogen present would have been "cooked" into heavier elements, in contradiction with the fact that about three-quarters of the present universe is hydrogen. This rapid nuclear cooking could have been prevented only if the universe was filled with radiation having an enormous equivalent temperature at very short wavelengths, which could blast nuclei apart as fast as they could be formed.

We are going to see that this radiation would have survived the subsequent expansion of the universe, but that its equivalent temperature would continue to fall as the universe expanded, in inverse proportion to the size of the universe. (As we shall see, this is essentially an effect of the red shift discussed in Chapter II.) It follows that the present universe should also be filled with radiation, but with an equivalent temperature vastly less than it was in the first few minutes. Peebles estimated that, in order for the radiation background to have kept the production of helium and heavier elements in the first few minutes within known bounds, it would have to have been so intense that its present temperature would be at least 10 degrees Kelvin.

The Cosmic Microwave Radiation Background

The figure of 10° K was somewhat of an overestimate, and this calculation was soon supplanted by more elaborate and accurate calculations by Peebles and others, which will be discussed in Chapter V. Peebles' preprint was in fact never published in its original form. However, the conclusion was substantially correct: from the observed abundance of hydrogen we can infer that the universe must in the first few minutes have been filled with an enormous amount of radiation which could prevent the formation of too much of the heavier elements; the expansion of the universe since then would have lowered its equivalent temperature to a few degrees Kelvin, so that it would appear now as a background of radio noise, coming equally from all directions. This immediately appeared as the natural explanation of the discovery of Penzias and Wilson. Thus, in a sense the antenna at Holmdel *is* in a box—the box is the whole universe. However, the equivalent temperature recorded by the antenna is not the temperature of the present universe, but rather the temperature that the universe had long ago, reduced in proportion to the enormous expansion that the universe has undergone since then.

Peebles' work was only the latest in a long series of similar cosmological speculations. In fact, in the late 1940s a "big bang" theory of nucleosynthesis had been developed by George Gamow and his collaborators, Ralph Alpher and Robert Herman, and was used in 1948 by Alpher and Herman to predict a radiation background with a present temperature of about 5° K. Similar calculations were carried out in 1964 by Ya. B. Zeldovich in Russia and independently by Fred Hoyle and R. J. Tayler in England. This earlier work was not at first known to the groups at Bell Laboratories and Princeton, and it did not have an effect on the actual discovery of the radiation background, so we may wait until Chapter VI to go into it in detail. We will also take up in Chapter VI the puzzling historical question of why none of this earlier theoretical work had led to a search for the cosmic microwave background.

Peebles' 1965 calculation had been instigated by the ideas of a senior experimental physicist at Princeton, Robert H. Dicke. (Among other things, Dicke had invented some of the key microwave techniques used by radio astronomers.) Sometime in 1964 Dicke had begun to wonder whether there might not be some observable radiation left over from a hot dense early stage of cosmic history. Dicke's speculations were based on an "oscillating" theory of the universe, to which we will return in the last chapter of this book. He apparently did not have a definite expectation of the temperature of this radiation, but he did appreciate the essential point, that there was something worth looking for. Dicke suggested to P. G. Roll and D. T. Wilkinson that they mount a search for a microwave radiation background, and they began to set up a small low-noise antenna on the roof of the Palmer Physical Laboratory at Princeton. (It is not necessary to use a large radio telescope for this purpose because the radiation comes from all directions, so that nothing is gained by having a more tightly focused antenna beam.)

Before Dicke, Roll, and Wilkinson could complete their measurements, Dicke received a call from Penzias, who had just heard from Burke of Peebles' work. They decided to publish a pair of companion letters in the *Astrophysical Journal*, in which Penzias and Wilson would announce their observations, and Dicke, Peebles, Roll, and Wilkinson would explain the cosmological interpretation. Penzias and Wilson, still very cautious, gave their paper the modest title "A Measurement of Excess Antenna Temperature at 4,080 Mc/s." (The frequency to which the antenna was tuned was 4,080 Mc/s, or 4,080

The Holmdel Radio Telescope: Arno Penzias (right) and Robert W. Wilson (left) are shown here with the 20-foot horn antenna used by them in 1964–65 in their discovery of the 3° K cosmic microwave radiation background. This telescope is at the Holmdel, New Jersey, site of the Bell Telephone Laboratories. (Bell Telephone Laboratories Photograph)

Inside the Holmdel Radio Telescope: Penzias is shown here taping the joints of the 20-fo[o]t horn antenna at Holmdel, with Wilson looking on. This was part of an effort to eliminate a possible source of electrical noise from the antenna structure that might account for the 3° microwave static observed in 1964–65. All such efforts only succeeded in reducing the observ[ed] microwave noise intensity very slightly, and the conclusion became inescapable that this microwa[ve] radiation is really of astronomical origin. (Bell Telephone Laboratories Photograph)

The Princeton Radio Antenna: This is a photograph of the original experiment at Princeton, which sought evidence of a cosmic radiation background. The small horn antenna is mounted facing upward on the wooden platform. Wilkinson is shown below the antenna and a little to the right; Roll, nearly hidden by the apparatus, is directly beneath the antenna. The shiny cylinder with a conical top is part of the cryogenic equipment used to maintain a liquid helium reference source whose radiation could be compared with that from the sky. This experiment confirmed the presence of a 3° K radiation background at a wavelength shorter than that used by Penzias and Wilson. (Princeton University Photograph)

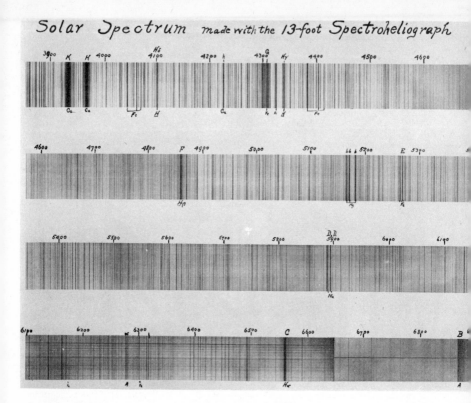

Solar Spectrum made with the 13-foot Spectroheliograph

The Spectrum of the Sun: This photograph shows light from the sun, broken up into its vario wavelengths by a 13-foot focus spectrograph. On the average, the intensity at different wav lengths is about the same as would be emitted by any totally opaque (or "black") body at temperature of 5800° K. However, the vertical dark "Fraunhofer" lines in the spectrum indica that light from the sun's surface is being absorbed by a relatively cool and partly transpare outer region, known as the reversing layer. Each dark line arises from the selective absorpti of light at one definite wavelength; the darker the line, the more intense the absorption. Wav lengths are indicated above the spectrum in Angstrom units (10^{-8}cm). Many of these lines a identified as due to absorption of light by specific elements, such as calcium (Ca), ir (Fe), hydrogen (H), magnesium (Mg), sodium (Na). It is partly through the study of su absorption lines that we can estimate cosmic abundances of the various chemical elemen Corresponding absorption lines in the spectra of distant galaxies are observed to be shift from their normal positions toward longer wavelengths; it is from this red shift that we in the expansion of the universe. (Hale Observatories Photograph) million cycles per second, corresponding to the wavelength of 7.35 centimeters.) They announced simply that "Measurements of the effective zenith noise temperature . . . have yielded a value of about 3.5° K higher than expected," and they avoided all mention of cosmology, except to note that "A possible explanation for the observed excess noise temperature is the one given by Dicke, Peebles, Roll, and Wilkinson in a companion letter in this issue."

Is the microwave radiation discovered by Penzias and Wilson actually left over from the beginning of the universe? Before we go on to consider the experiments that have been performed since 1965 to settle this question, it will be necessary for us first to ask what we expect theoretically: What are the general properties of the radiation that *should* be filling the universe if current cosmological ideas are correct? This question leads us to consider what happens to radiation as the universe expands—not only at the time of nucleosynthesis, at the end of the first three minutes, but in the aeons that have elapsed since then.

It will be very helpful here if we now give up the classical picture of radiation in terms of electromagnetic waves that we have been using up to this point, and adopt instead the more modern "quantum" view that radiation consists of particles, known as *photons*. An ordinary light wave contains a huge number of photons traveling along together, but if we were to measure the energy carried by the train of waves very precisely, we would find that it is always some multiple of a definite quantity, which we identify as the energy of a single photon. As we shall see, photon energies are generally quite small, so that for most practical purposes it appears as if an electromagnetic wave could have any energy whatever. However, the interaction of radiation with atoms or atomic nuclei usually takes place one photon at a time, and in studying such processes it is necessary to adopt a photon rather than a wave description. Photons have zero mass and zero electrical charge, but they are real nonetheless—each one carries a definite energy and momentum, and even has a definite spin around its direction of motion.

What happens to an individual photon as it travels along through the universe? Not much, as far as the present universe is concerned. The light from objects some 10,000 million light years away seems to reach us perfectly well. Thus whatever matter may be present in intergalactic space must be sufficiently transparent so that photons can travel for an appreciable fraction of the age of the universe without being scattered or absorbed.

However, the red shifts of the distant galaxies tell us that the universe is expanding, so its contents must once have been much more compressed than now. The temperature of a fluid will generally rise when the fluid is compressed, so we can also infer that the matter of the universe was much hotter in the past. We believe in fact that there was a time, which as we shall see lasted perhaps for the first 700,000 years of the universe, when the contents of the universe were so hot and dense that they could not yet have clumped into stars and galaxies, and even the atoms were still broken up into their constituent nuclei and electrons.

Under these unpleasant conditions a photon could not travel immense distances without hindrance, as it can in our present universe. A photon would find in its path a huge number of free electrons which could efficiently scatter or absorb it. If the photon is scattered by an electron it will generally either lose a little energy to the electron or gain a little energy from it, depending on whether the proton initially has more or less energy than the electron. The "mean free time" that the photon could travel before it was absorbed or suffered an appreciable change in energy would have been quite short, much shorter than the characteristic time of the expansion of the universe. The corresponding mean free times for the other particles, the electrons and atomic nuclei, would have been even shorter. Thus, although in a sense the universe was expanding very rapidly at first, to an individual photon or electron or nucleus the expansion was taking plenty of time, time enough for each particle to be scattered or absorbed or reemitted many times as the universe expanded.

Any system of this sort, in which the individual particles have time for many interactions, is expected to come to a state of equilibrium. The numbers of particles with properties (position, energy, velocity, spin, and so on) in a certain range will settle down to a value such that an equal number of particles are knocked out of the range every second as are knocked into it. Thus, the properties of such a system will not be determined by any initial conditions, but rather by the requirement that the equilibrium be maintained. Of course, "equilibrium" here does not mean that the particles are frozen—each one is continually being knocked about by its neighbors. Rather, the equilibrium is statistical—it is the way that the particles are distributed in position, energy, and so on, that does not change, or changes slowly.

Equilibrium of this statistical kind is usually known as "thermal equilibrium," because a state of equilibrium of this kind is always characterized by a definite temperature which must be uniform throughout the system. Indeed, strictly speaking, it is only in a state of thermal equilibrium that temperature can be precisely defined. The powerful and profound branch of theoretical physics known as "statistical mechanics" provides a mathematical machinery for computing the properties of any system in thermal equilibrium.

The approach to thermal equilibrium works a little like the way the price mechanism is supposed to work in classical economics. If demand exceeds supply, the price of goods will rise, cutting the effective demand and encouraging increased production. If supply exceeds demand, prices will drop, increasing effective demand and discouraging further production. In either case, supply and demand will approach equality. In the same way, if there are too many or too few particles with energies, velocities, and so on, in some particular range, then the rate at which they leave this range will be greater or less than the rate at which they enter, until equilibrium is established.

Of course, the price mechanism does not always work exactly the way it is supposed to in classical economics, but here too the analogy holds—most physical systems in the real world are quite far from thermal equilibrium. At the centers of stars there is nearly perfect thermal equilibrium, so we can estimate what conditions are like there with some confidence, but the surface of the earth is nowhere near equilibrium, and we cannot be sure whether or not it will rain tomorrow. The universe has never been in perfect thermal equilibrium, because after all it *is* expanding. However, during the early period when the rates of scattering or absorption of individual particles were much faster than the rate of cosmic expansion, the universe could be regarded as evolving "slowly" from one state of nearly perfect thermal equilibrium to another.

It is crucial to the argument of this book that the universe has once passed through a state of thermal equilibrium. According to the conclusions of statistical mechanics, the properties of any system in thermal equilibrium are entirely determined once we specify the temperature of the system and the densities of a few conserved quantities (about which more in the next chapter). Thus, the universe preserves only a very limited memory of its initial conditions. This is a pity, if what we want is to reconstruct the very beginning, but it also offers a compensation in that we can infer the course of events since the beginning without too many arbitrary assumptions.

We have seen that the microwave radiation discovered by Penzias and Wilson is believed to be left over from a time when the universe was in a state of thermal equilibrium. Therefore, in order to see what properties we expect for the observed microwave radiation background, we have to ask: What are the general properties of radiation in thermal equilibrium with matter?

As it happens, this is precisely the question which historically gave rise to the quantum theory and the interpretation of radiation in terms of photons. By the 1890s it had become known that the properties of radiation in a state of thermal equilibrium with matter depend only on the temperature. To be more specific, the amount of energy per unit volume in such radiation within any given range of wavelengths is given by a universal formula, involving only the wavelength and the temperature. The same formula gives the amount of radiation inside a box with opaque walls, so a radio astronomer can use this formula to interpret the intensity of the radio noise he observes in terms of an "equivalent temperature." Essentially the same formula also gives the amount of radiation emitted per second and per square centimeter at any wavelength from any totally absorbing surface, so radiation of this sort is generally known as "black-body radiation." That is, black-body radiation is characterized by a definite distribution of energy with wavelength, given by a universal formula depending only on the temperature. The hottest question facing the theoretical physicists of the 1890s was to find this formula.

The correct formula for black-body radiation was found in the closing weeks of the nineteenth century by Max Karl Ernst Ludwig Planck. The precise form of Planck's result is shown in figure 7, for the particular temperature 3° K of the observed cosmic microwave noise. Planck's formula can be summarized qualitatively as follows: In a box filled with black-body radiation, the energy in any range of wavelengths rises very steeply with increasing wavelength, reaches a maximum, and then falls off steeply again. This "Planck distribution" is universal, not dependent on the nature of the matter with which the radiation interacts, but only on its temperature. As used today, the term "black-body radiation" means any radiation in which the distribution of energy with wavelength matches the Planck formula, whether or not the radiation was actually emitted by a black body. Thus, during at least the first million years or so, when radiation and matter were in thermal equilibrium, the universe must have been filled with black-body radiation with a temperature equal to that of the material contents of the universe.

The importance of Planck's calculation went far beyond the problem of black-body radiation, because in it he introduced a

Figure 7. *The Planck Distribution.* The energy density per unit wavelength range is shown as a function of wavelength, for black-body radiation with a temperature of 3° K. (For a temperature which is greater than 3° K by a factor f, it is only necessary to reduce the wavelengths by a factor $1/f$ and increase the energy densities by a factor f^5.) The straight part of the curve on the right is approximately described by the simpler "Rayleigh-Jeans distribution"; a line with this slope is expected for a wide variety of cases besides black-body radiation. The steep falloff to the left is due to the quantum nature of radiation, and is a specific feature of black-body radiation. The line marked "galactic radiation" shows the intensity of radio noise from our galaxy. (Arrows indicate the wavelength of the original Penzias and Wilson measurement, and the wavelength at which a radiation temperature could be inferred from measurements of absorption by the first excited rotational state of interstellar cyanogen.)

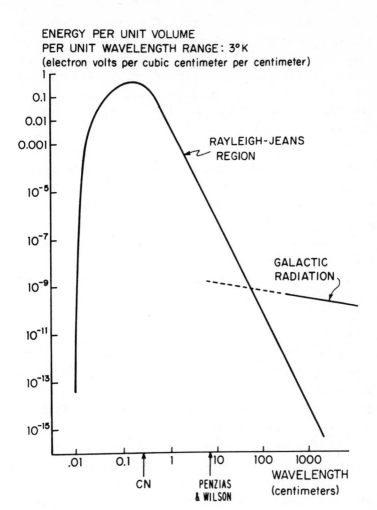

new idea, that energies come in distinct chunks, or "quanta." Planck originally considered only the quantization of the energy of the matter in equilibrium with radiation, but Einstein suggested a few years later that radiation itself comes in quanta, later called photons. These developments eventually led in the 1920s to one of the great intellectual revolutions in the history of science, the replacement of classical mechanics by an entirely new language, that of quantum mechanics.

We are not going to be able to go far into quantum mechanics in this book. However, it will help us in understanding the behavior of radiation in an expanding universe to take a look at how the picture of radiation in terms of photons leads to the general features of the Planck distribution.

The reason that the energy density of black-body radiation falls off for very large wavelengths is simple: It is hard to fit radiation into any volume whose dimensions are smaller than the wavelength. This much could be (and was) understood without the quantum theory, simply on the basis of the older wave theory of radiation.

On the other hand, the decrease of the energy density of black-body radiation for very short wavelengths could not be understood in a nonquantum picture of radiation. It is a well-known consequence of statistical mechanics that at any given temperature it is difficult to produce any kind of particle or wave or other excitation whose energy is greater than a certain definite amount, proportional to the temperature. However, if wavelets of radiation could have arbitrarily small energies, then there would be nothing to limit the total amount of black-body radiation of very short wavelengths. Not only was this in contradiction with experiment—it would have led to the catastrophic result of the total energy of black-body radiation being infinite! The only way out was to suppose that the energy comes in chunks or "quanta," with the amount of energy in each chunk increasing with decreasing wavelength, so that at any given temperature there would be very little radiation at the short wavelengths for which the chunks are highly energetic. In the final formulation of this hypothesis due to Einstein, *the energy of any photon is inversely proportional to the wavelength*; at any given temperature, black-body radiation will contain very few photons that have too large an energy, and therefore very few that have too short a wavelength, thus explaining the falloff of the Planck distribution at short wavelengths.

To be specific, the energy of a photon with a wavelength of one centimeter is 0.000124 electron volts, and proportionally more at shorter wavelengths. The electron volt is a convenient unit of energy, equal to the energy gained by one electron in moving across a voltage drop of one volt. For instance, an ordinary 1.5 volt flashlight battery expends 1.5 electron volts for every electron that it pushes through the filament of the light bulb. (In terms of the metric units of energy, one electron volt is 1.602×10^{-12} ergs, or 1.602×10^{-19} joules.) According to Einstein's rule, the energy of a photon at the 7.35 centimeter microwave wavelength to which Penzias and Wilson were tuned was 0.000124 electron volts divided by 7.35, or 0.000017 electron volts. On the other hand, a typical photon in visible light would have a wavelength of about a twenty-thousandth of a centimeter (5×10^{-5} cm), so its energy would be 0.000124 electron volts times 20,000, or about 2.5 electron volts. In either case the energy of a photon is very small in macroscopic terms, which is why photons seem to blend together into continuous streams of radiation.

Incidentally, chemical reaction energies are generally of the order of an electron volt per atom or per electron. For instance, to rip the electron out of a hydrogen atom altogether takes 13.6 electron volts, but this is an exceptionally violent chemical event. The fact that photons in sunlight also have energies of the order of an electron volt or so is tremendously important to us; it is what allows these photons to produce chemical reactions essential to life, such as photosynthesis. Nuclear reaction energies are generally of the order of a *million* electron volts per atomic nucleus, which is why a pound of plutonium has roughly the explosive energy of a million pounds of TNT.

The photon picture allows us easily to understand the chief qualitative properties of black-body radiation. First, the principles of statistical mechanics tell us that the typical photon energy is proportional to the temperature, while Einstein's rule tells us that any photon's wavelength is inversely proportional to the photon energy. Hence, puting these two rules together, the typical wavelength of photons in black-body radiation is inversely proportional to the temperature. To put it quantitatively, the typical wavelength near which most of the energy of black-body radiation is concentrated is 0.29 centimeters at a temperature of 1° K, and proportionally less at higher temperatures.

For instance, an opaque body at an ordinary "room" temperature of 300° K (= 27°C) will emit black-body radiation with a typical wavelength of 0.29 centimeters divided by 300, or about a thousandth of a centimeter. This is in the range of infrared radiation, and is too long a wavelength for our eyes to see. On the other hand, the surface of the sun is at a temperature of about 5,800° K, and in consequence the light it emits is peaked at a wavelength of about 0.29 centimeters divided by 5,800, that is, about five hundred-thousandths of a centimeter $(5 \times 10^{-5}$ cm) or, equivalently, about 5,000 Angstrom units.

(One Angstrom unit is one hundred-millionth or 10^{-8} of a centimeter.) As already mentioned, this is in the middle of the range of wavelengths that our eyes evolved to be able to see, and which we call "visible" wavelengths. The fact that these wavelengths are so short explains why it was not until the beginning of the nineteenth century that light was discovered to have a wave nature; it is only when we examine the light that passes through really small holes that we can notice phenomena characteristic of wave propagation, such as diffraction.

We also saw that the decrease in the energy density of black-body radiation at long wavelengths is due to the difficulty of putting radiation in any volume whose dimensions are smaller than a wavelength. In fact, the average distance between photons in black-body radiation is roughly equal to the typical photon wavelength. But we saw that this typical wavelength is inversely proportional to the temperature, so the average distance between photons is also inversely proportional to the temperature. The number of things of any kind in a fixed volume is inversely proportional to the cube of their average separation, so in black-body radiation the rule is that *the number of photons in a given volume is proportional to the cube of the temperature.*

We can put together this information to draw some conclusions about the amount of energy in black-body radiation. The energy per liter, or "energy density," is simply the number of photons per liter times the average energy per photon. But we have seen that the number of photons per liter is proportional to the cube of the temperature, while the average photon energy is simply proportional to the temperature. Hence the energy per liter in black-body radiation is proportional to the cube of the temperature times the temperature, or, in other

63

words, to the *fourth* power of the temperature. To put it quantitatively, the energy density of black-body radiation is 4.72 electron volts per liter at a temperature of 1° K, 47,200 electron volts per liter at a temperature of 10° K, and so on. (This is known as the Stefan-Boltzmann law.) If the microwave noise discovered by Penzias and Wilson really is black-body radiation with a temperature of 3° K, then its energy density must be 4.72 electron volts per liter times 3 to the fourth power, or about 380 electron volts per liter. When the temperature was a thousand times larger, the energy density was a million million (10^{12}) times larger.

Now we can return to the origin of the fossil microwave radiation. We have seen that there must have been a time when the universe was so hot and dense that atoms were dissociated into their nuclei and electrons, and the scattering of photons by free electrons maintained a thermal equilibrium between matter and radiation. As time passed, the universe expanded and cooled, eventually reaching a temperature (about 3,000° K) cool enough to allow the combination of nuclei and electrons into atoms. (In the astrophysical literature this is usually called "recombination," a singularly inappropriate term, for at the time we are considering the nuclei and electrons had never in the previous history of the universe been combined into atoms!) The sudden disappearance of free electrons broke the thermal contact between radiation and matter, and the radiation continued thereafter to expand freely.

At the moment this happened, the energy in the radiation field at various wavelengths was governed by the conditions of thermal equilibrium, and was therefore given by the Planck black-body formula for a temperature equal to that of the matter, about 3,000° K. In particular, the typical photon wavelength would have been about one micron (a ten-thousandth of a centimeter, or 10,000 Angstroms) and the average distance between photons would have been roughly equal to this typical wavelength.

What has happened to the photons since then? Individual photons would not be created or destroyed, so the average distance between photons would simply increase in proportion to the size of the universe, i.e., in proportion to the average distance between typical galaxies. But we saw in the last chapter that the effect of the cosmological red shift is to "pull out" the wavelength of any ray of light as the universe expands; thus, the wavelengths of any individual photon would also simply increase in proportion to the size of the universe. The photons would therefore remain about one typical wavelength apart, just as for black-body radiation. Indeed, by pursuing this line of argument quantitatively, one can show that *the radiation filling the universe would continue to be described precisely by the Planck black-body formula as the universe expanded,* even though it was no longer in thermal equilibrium with the matter. (See mathematical note 4, p. 173.) The only effect of the expansion is to increase the typical photon wavelength in proportion to the size of the universe. The temperature of the black-body radiation is inversely proportional to the typical wavelength, so it would fall as the universe expanded, in inverse proportion to the size of the universe.

For instance, Penzias and Wilson found that the intensity of the microwave static they had discovered corresponded to a temperature of roughly 3° K. This is just what would be expected if the universe has expanded by a factor of 1,000 since the time when the temperature was high enough (3,000° K) to keep matter and radiation in thermal equilibrium. If this interpretation is correct, the 3° K radio static is by far the most ancient signal received by astronomers, having been emitted long before the light from the most distant galaxies that we can see.

But Penzias and Wilson had measured the intensity of the cosmic radio static at only one wavelength, 7.35 centimeters. Immediately it became a matter of extreme urgency to decide whether the distribution of radiant energy with wavelength is described by the Planck black-body formula, as would be expected if this really were red-shifted fossil radiation left over from some epoch when the radiation and matter of the universe were in thermal equilibrium. If so, then the "equivalent temperature," calculated by matching the observed radio noise intensity to the Planck formula, should have the same value at all wavelengths as at the 7.35 centimeter wavelength studied by Penzias and Wilson.

As we have seen, at the time of the discovery by Penzias and Wilson there already was another effort under way in New Jersey to detect a cosmic microwave radiation background. Soon after the original pair of papers by the Bell Laboratories and Princeton groups, Roll and Wilkinson announced their own result: the equivalent temperature of the radiation background at a wavelength of 3.2 centimeters was between 2.5 and 3.5 degrees Kelvin. That is, within experimental error, the intensity of the cosmic static at 3.2 centimeters wavelength was greater than at 7.35 centimeters by just the ratio that would be expected if the radiation is described by the Planck formula!

Since 1965 the intensity of the fossil microwave radiation has been measured by radio astronomers at over a dozen wavelengths ranging from 73.5 centimeters down to 0.33 centimeters. Every one of these measurements is consistent with a Planck distribution of energy versus wavelength, with a temperature between 2.7° K and 3° K.

However, before we jump to the conclusion that this really is black-body radiation, we should recall that the "typical" wavelength, at which the Planck distribution reaches its maximum, is 0.29 centimeters divided by the temperature in degrees Kelvin, which for a temperature of 3° K works out to just under 0.1 centimeter. Thus all these microwave measurements have been on the *long* wavelength side of the maximum in the Planck distribution. But we have seen that the increase in energy density with decreasing wavelength in this part of the spectrum is just due to the difficulty of putting large wavelengths in small volumes, and would be expected for a wide variety of radiation fields, including radiation that was *not* produced under conditions of thermal equilibrium. (Radio astronomers refer to this part of the spectrum as the Rayleigh-Jeans region, because it was first analyzed by Lord Rayleigh and Sir James Jeans.) In order to verify that we really are seeing black-body radiation, it is necessary to go beyond the maximum of the Planck distribution into the short-wavelength region, and check that the energy density really does fall off with decreasing wavelength, as expected on the basis of the quantum theory. At wavelengths shorter than 0.1 centimeter we are really outside the realm of the radio or microwave astronomers, and into the newer discipline of infrared astronomy.

Unfortunately the atmosphere of our planet, which is nearly transparent at wavelengths above 0.3 centimeters, becomes increasingly opaque at shorter wavelengths. It does not seem likely that any ground-based radio observatory, even one located at mountain altitude, will be able to measure the cosmic radiation background at wavelengths much shorter than 0.3 centimeters.

Oddly enough, the radiation background *was* measured at

shorter wavelengths, long before any of the astronomical work discussed so far in this chapter, and by an optical rather than by a radio or infrared astronomer! In the constellation Ophiuchus ("the serpent bearer") there is a cloud of interstellar gas which happens to lie between the earth and a hot but otherwise unremarkable star, ζ Oph. The spectrum of ζ Oph is crossed with a number of unusual dark bands, indicating that the intervening gas is absorbing light at a set of sharp wavelengths. These are the wavelengths at which photons have just the energies required to induce transitions in the molecules of the gas cloud, from states of lower to states of higher energy. (Molecules, like atoms, exist only in states of distinct, or "quantized," energy.) Thus, observing the wavelengths where the dark bands occur, it is possible to infer something about the nature of these molecules, and of the states in which they are found.

One of the absorption lines in the spectrum of ζ Oph is at a wavelength of 3,875 Angstrom units (38.75 millionths of a centimeter), indicating the presence in the interstellar cloud of a molecule, cyanogen (CN), consisting of one carbon and one nitrogen atom. (Strictly speaking, CN should be called a "radical," meaning that under normal conditions it combines rapidly with other atoms to form more stable molecules, such as the poison, hydrocyanic acid (HCN). In interstellar space CN is quite stable.) In 1941 it was found by W. S. Adams and A. McKellar that this absorption line is actually split, consisting of three components with wavelengths 3874.608 Angstroms, 3875.763 Angstroms, and 3873.998 Angstroms. The first of these absorption wavelengths corresponds to a transition in which the cyanogen molecule is lifted from its state of lowest energy (the "ground state") to a *vibrating* state, and would be expected to be produced even if the cyanogen

68

were at zero temperature. However, the other two lines could only be produced by transitions in which the molecule is lifted from a *rotating* state just above the ground state to various other vibrating states. Thus, a fair fraction of the cyanogen molecules in the interstellar cloud must be in this rotating state. Using the known energy difference between the ground state and the rotating state, and the observed relative intensities of the various absorption lines, McKellar was able to estimate that the cyanogen was being exposed to some sort of perturbation with an effective temperature of about 2.3° K, which could lift the cyanogen molecule into the rotating state.

At the time there did not seem to be any reason to associate this mysterious perturbation with the origin of the universe, and it did not receive a great deal of attention. However, after the discovery of a 3° K cosmic radiation background in 1965, it was realized (by George Field, I. S. Shklovsky, and N. J. Woolf) that this was just the perturbation that had been observed in 1941 to be producing the rotation of the cyanogen molecules in the Ophiuchus clouds. The wavelength of the black-body photons which would be needed to produce this rotation is 0.263 centimeters, shorter than any wavelength accessible to ground-based radio astronomy, but still not short enough to test the rapid falloff of wavelengths below 0.1 cm expected for a 3° K Planck distribution.

Since then there has been a search for other absorption lines caused by excitation of cyanogen molecules in other rotating states, or of other molecules in various rotating states. The observation in 1974 of absorption by the second rotating state of interstellar cyanogen has yielded an estimate of the radiation intensity at a wavelength of 0.132 centimeters, also corresponding to a temperature of about 3° K. However, such observations have so far set only upper limits on the radiation energy density at wavelengths shorter than 0.1 centimeter. These results are encouraging because they indicate that the radiation energy density does begin to fall off steeply at some wavelength around 0.1 centimeters, as expected if this is black-body radiation. However, these upper limits do not allow us to verify that this really is black-body radiation, or to determine a precise radiation temperature.

It has only been possible to attack this problem by lifting an infrared receiver above the earth's atmosphere, either with a balloon or a rocket. These experiments are extraordinarily difficult and at first gave inconsistent results, alternately encouraging either the adherents of the standard cosmology or its opponents. A Cornell rocket group found much more radiation at short wavelengths than could be expected for a Planck black-body distribution, while an M.I.T. balloon group obtained results roughly consistent with those expected for black-body radiation. Both groups continued their work, and by 1972 they were both reporting results indicating a black-body distribution with temperature close to 3° K. In 1976 a Berkeley balloon group confirmed that the radiation energy density continues to fall off for short wavelengths in the range of 0.25 centimeters to 0.06 centimeters, in the manner expected for a temperature within 0.1° K of 3° K. It now seems to be settled that the cosmic radiation background really is black-body radiation, with a temperature close to 3° K.

The reader may be wondering at this point why this question could not have been settled by simply mounting infrared equipment in an artificial earth satellite, and taking all the time needed to make accurate measurements well above the earth's atmosphere. I am not really sure why this has been impossible. The reason usually given is that in order to measure radiation temperatures as low as 3° K it is necessary to cool the apparatus with liquid helium (a "cold load"), and there does not exist a technology for carrying this sort of cryogenic equipment aboard an earth satellite. However, one cannot help suspecting that these truly cosmic investigations deserve a larger share of the space budget.

The importance of carrying out observations above the earth's atmosphere appears even greater when we consider the distribution of the cosmic radiation background with *direction* as well as with wavelength. All observations so far are consistent with a radiation background that is perfectly isotropic, i.e., independent of direction. As mentioned in the preceding chapter, this is one of the most powerful arguments in favor of the Cosmological Principle. However, it is very difficult to distinguish a possible direction dependence that is intrinsic to the cosmic radiation background from one that is merely due to effects of the earth's atmosphere; indeed, in measurements of the radiation background temperature, the radiation background is distinguished from the radiation from our atmosphere by *assuming* that it is isotropic.

The thing that makes the direction dependence of the microwave radiation background such a fascinating subject for study is that the intensity of this radiation is not expected to be perfectly isotropic. There might be fluctuations in the intensity with small changes in direction, caused by the actual lumpiness of the universe either at the time the radiation was emitted or since then. For instance, galaxies in the first stages of formation might show up as warm spots in the sky, with slightly higher black-body temperature than average, extending perhaps over half a minute of arc. In addition, there almost certainly is a small smooth variation of the radiation intensity around the whole sky, caused by the earth's motion through the universe. The earth is going around the sun at a speed of 30 kilometers per second, and the solar system is being carried along by the rotation of our galaxy at a speed of about 250 kilometers per second. No one knows precisely what velocity our galaxy has relative to the cosmic distribution of typical galaxies, but presumably it moves at a few hundred kilometers per second in some direction. If, for example, we suppose that the earth is moving at a speed of 300 kilometers per second relative to the average matter of the universe, and hence relative to the radiation background, then the wavelength of the radiation coming from ahead or astern of the earth's motion should be decreased or increased, respectively, by the ratio of 300 kilometers per second to the speed of light, or 0.1 percent. Thus, the equivalent radiation temperature should vary smoothly with direction, being about 0.1 percent higher than average in the direction toward which the earth is going and about 0.1 percent lower than average in the direction from which we have come. For the last few years the best upper limit on any direction dependence of the equivalent radiation temperature has been just about 0.1 percent, so we have been in the tantalizing position of being almost but not quite able to measure the velocity of the earth through the universe. It may not be possible to settle this question until measurements can be made from satellites orbiting the earth. (As final corrections were being made in this book I received a *Cosmic Background Explorer Satellite Newsletter* #1 from John Mather of N.A.S.A. It announces the appointment of a team of six scientists, under Rainier Weiss of M.I.T., to study the possible measurement of the infrared and microwave radiation backgrounds from space. Bon voyage.)

We have observed that the cosmic microwave radiation background provides powerful evidence that the radiation and matter of the universe were once in a state of thermal equilibrium. However, we have not yet drawn much cosmological insight from the particular observed numerical value of the equivalent radiation temperature, 3° K. In fact, this radiation temperature allows us to determine the one crucial number that we will need to follow the history of the first three minutes.

As we have seen, at any given temperature, the number of photons per unit volume is inversely proportional to the cube of a typical wavelength, and hence directly proportional to the cube of the temperature. For a temperature of precisely 1° K there would be 20,282.9 photons per liter, so the 3° K radiation background contains about 550,000 photons per liter. However, the density of nuclear particles (neutrons and protons) in the present universe is somewhere between 6 and 0.03 particles per *thousand* liters. (The upper limit is twice the critical density discussed in Chapter II; the lower limit is a low estimate of the density actually observed in visible galaxies.) Thus, depending on the actual value of the particle density, there are between 100 million and 20,000 million photons for every nuclear particle in the universe today.

Furthermore, this enormous ratio of photons to nuclear particles has been roughly constant for a very long time. During the period that the radiation has been expanding freely (since the temperature dropped below about 3,000° K) the background photons and the nuclear particles have been neither created nor destroyed, so their ratio has naturally remained constant. We will see in the next chapter that this ratio was roughly constant even earlier, when individual photons *were* being created and destroyed.

This is the most important quantitative conclusion to be drawn from measurements of the microwave radiation background—as far back as we can look in the early history of the universe there have been between 100 million and 20,000 million photons per neutron or proton. In order not to sound unnecessarily equivocal, I will round off this number in what follows, and suppose for purposes of illustration that there are now and have been just 1,000 million photons per nuclear particle in the average contents of the universe.

One very important consequence of this conclusion is that the differentiation of matter into galaxies and stars could not have begun until the time when the cosmic temperature became low enough for electrons to be captured into atoms. In order for gravitation to produce the clumping of matter into isolated fragments that had been envisioned by Newton, it is necessary for gravitation to overcome the pressure of matter and the associated radiation. The gravitational force within any nascent clump increases with the size of the clump, while the pressure does not depend on the size; hence at any given density and pressure, there is a minimum mass which is susceptible to gravitational clumping. This is known as the "Jeans mass," because it was first introduced in theories of the formation of stars by Sir James Jeans in 1902. It turns out that the Jeans mass is proportional to the three-halves power of the pressure (see mathematical note 5, p. 174). Just before the electrons started to be captured into atoms, at a temperature of about 3,000° K, the pressure of radiation was enormous, and the Jeans mass was correspondingly large, about a million or so times larger than the mass of a large galaxy. Neither galaxies nor even clusters of galaxies are large enough to have formed at this time. However, a little later the electrons joined with nuclei into atoms; with the disappearance of free electrons, the universe became transparent to radiation; and so the radiation pressure became ineffective. At a given temperature and density the pressure of matter or radiation is simply proportional to the number of particles or photons, respectively, so when the radiation pressure became ineffective the total effective pressure dropped by a factor of about 1,000 million. The Jeans mass dropped by the three-halves power of this factor, to about one-millionth the mass of a galaxy. From then on the pressure of matter alone would be far too weak to resist the clumping of matter into the galaxies we see in the sky.

This is not to say that we actually understand how galaxies are formed. The theory of the formation of galaxies is one of the great outstanding problems of astrophysics, a problem that today seems far from solution. But that is another story. For us, the important point is that in the early universe, at temperatures above about 3,000° K, the universe consisted not of the galaxies and stars we see in the sky today, but only of an ionized and undifferentiated soup of matter and radiation.

Another remarkable consequence of the huge ratio of photons to nuclear particles is that there must have been a time, not relatively far in the past, when the energy of radiation was greater than the energy contained in the matter of the universe. The energy in the mass of a nuclear particle is given by Einstein's formula $E = mc^2$ as about 939 million electron volts. The average energy of a photon in 3° K black-body radiation is very much less, about 0.0007 electron volts, so that even with 1,000 million photons per neutron or proton, most of the energy of the present universe is in the form of matter, not radiation. However, at earlier times the temperature was higher, so the energy of each photon was higher, while the energy in a neutron or proton mass was always the same. With 1,000 million photons per nuclear particle, in order for the radiation energy to exceed the energy of matter it is only necessary that the mean energy of a black-body photon be greater

75

than about one thousand-millionth of the energy of a nuclear particle mass, or about one electron volt. This was the case when the temperature was about 1,300 times greater than at present, or about 4,000° K. This temperature marks the transition between a "radiation-dominated" era, in which most of the energy in the universe was in the form of radiation, and the present "matter-dominated" era, in which most of the energy is in the masses of the nuclear particles.

It is striking that the transition from a radiation- to a matter-dominated universe occurred at just about the same time that the contents of the universe were becoming transparent to radiation, at about 3,000° K. No one really knows why this should be so, although there have been interesting suggestions. We also do not really know which transition occurred first: If there were now 10,000 million photons per nuclear particle, then radiation would have continued to predominate over matter until the temperature dropped to 400° K, well after the contents of the universe became transparent.

These uncertainties will not interfere with our story of the early universe. The important point for us is that at any time well before the contents of the universe became transparent, the universe could be regarded as composed chiefly of radiation, with only a small contamination of matter. The enormous energy density of radiation in the early universe has been lost by the shift of photon wavelengths to the red as the universe expanded, leaving the contamination of nuclear particles and electrons to grow into the stars and rocks and living beings of the present universe.

IV
_____

RECIPE FOR
A HOT UNIVERSE

T HE OBSERVATIONS discussed in the last two
chapters have revealed that the universe is expanding, and that
it is filled with a universal background of radiation, now at a
temperature of about 3° K. This radiation appears to be left
over from a time when the universe was effectively opaque,
when it was about 1,000 times smaller and hotter than at
present. (As always, when we speak of the universe being
1,000 times smaller than at present we mean simply that the
distance between any given pair of typical particles was 1,000
times less then than now.) As a final preparation for our account of the first three minutes we must look back to yet earlier
times, when the universe was even smaller and hotter, using
the eye of theory rather than optical or radio telescopes to examine the physical conditions that prevailed.

D

At the end of Chapter III we noted that when the universe was 1,000 times smaller than at present, and its material contents were just on the verge of becoming transparent to radiation, the universe was also passing from a radiation-dominated era to the present matter-dominated era. During the radiation-dominated era there was not only the same enormous number of photons per nuclear particle that exists now, but the energy of the individual photons was sufficiently high so that most of the energy of the universe was in the form of radiation, not mass. (Recall that photons are the massless particles, or "quanta," of which light, according to the quantum theory, is composed.) Hence, it should be a good approximation to treat the universe during this era as if it were filled purely with radiation, with essentially no matter at all.

One important qualification has to be attached to this conclusion. We will see in this chapter that the age of pure radiation actually began only at the end of the first few minutes, when the temperature had dropped below a few thousand million degrees Kelvin. At earlier times matter *was* important, but matter of a kind very different from that of which our present universe is composed. However, before we look that far back, let us first consider briefly the true era of radiation, from the end of the first few minutes up to the time, a few hundred thousand years later, when matter again became more important than radiation.

In order to follow the history of the universe during this era, all we need to know is how hot everything was at any given moment. Or to put it a different way—how is the temperature related to the size of the universe as the universe expands?

It would be easy to answer this question if the radiation could be considered to be expanding freely. The wavelength of each photon would have simply been stretched out (by the red shift) in proportion to the size of the universe, as the universe expanded. Furthermore, we have seen in the preceding chapter that the average wavelength of black-body radiation is inversely proportional to its temperature. Thus the temperature would have decreased in inverse proportion to the size of the universe, just as it is doing right now.

Fortunately for the theoretical cosmologist, the same simple relation holds even when we take into account the fact that the radiation was not really expanding freely—rapid collisions of photons with the relatively small number of electrons and nuclear particles made the contents of the universe opaque during the radiation-dominated era. While a photon was in free flight between collisions, its wavelength would have increased in proportion to the size of the universe, and there were so many photons per particle that the collisions simply forced the matter temperature to follow the radiation temperature, not vice versa. Thus, for instance, when the universe was ten thousand times smaller than now, the temperature would have been proportionally higher than now, or about 30,000° K. So much for the true era of radiation.

Eventually, as we look farther and farther back into the history of the universe, we come to a time when the temperature was so high that collisions of photons with each other could produce material particles out of pure energy. We are going to find that the particles produced in this way out of pure radiant energy were just as important during the first few minutes as radiation, both in determining the rates of various nuclear reactions and in determining the rate of expansion of the universe itself. Therefore, in order to follow the course of events at really early times, we are going to need to know how hot the universe had to be to produce large numbers of material particles out of the energy of radiation, and how many particles were thus produced.

The process by which matter is produced out of radiation

can best be understood in terms of the quantum picture of light. Two quanta of radiation, or photons, may collide and disappear, all their energy and momentum going into the production of two or more material particles. (This process is actually observed indirectly in present-day high-energy nuclear physics laboratories.) However, Einstein's Special Theory of Relativity tells us that a material particle even at rest will have a certain "rest energy" given by the famous formula $E = mc^2$. (Here c is the speed of light. This is the source of the energy released in nuclear reactions, in which a fraction of the mass of atomic nuclei is annihilated.) Hence, in order for two photons to produce two material particles of mass m in a head-on collision, the energy of each photon must be at least equal to the rest energy mc^2 of each particle. The reaction will still occur if the energy of the individual photons is greater than mc^2; the extra energy will simply go into giving the material particles a high velocity. However, particles of mass m cannot be produced in collisions of two photons if the energy of the photons is below mc^2, because there is then not enough energy to produce even the mass of these particular particles.

Evidently, in order to judge the effectiveness of radiation in producing material particles, we have to know the characteristic energy of the individual photons in the radiation field. This can be estimated well enough for our present purposes by using a simple rule of thumb: to find the characteristic photon energy, simply multiply the temperature of the radiation by a fundamental constant of statistical mechanics, known as Boltzmann's constant. (Ludwig Boltzmann was, along with the American Willard Gibbs, the founder of modern statistical mechanics. His suicide in 1906 is said to be due at least in part to philosophical opposition to his work, but all these controversies are long settled.) The value of Boltzmann's constant is 0.00008617 electron volts per degree Kelvin. For instance, at the temperature of 3,000° K, when the contents of the universe were just becoming transparent, the characteristic energy of each photon was about equal to 3,000° K times Boltzmann's constant, or 0.26 electron volts. (Recall that an electron volt is the energy acquired by one electron in moving through an electrical potential difference of one volt. Chemical reaction energies are typically of the order of an electron volt per atom; this is why radiation at temperatures above 3,000° K is hot enough to keep a significant fraction of electrons from being incorporated into atoms.)

We saw that in order to produce material particles of mass m in collisions of photons, the characteristic photon energy has to be at least equal to the energy mc^2 of the particles at rest. Since the characteristic photon energy is the temperature times Boltzmann's constant, it follows that the temperature of the radiation has to be at least of the order of the rest energy mc^2 divided by Boltzmann's constant. That is, for each type of material particle there is a "threshold temperature," given by the rest energy mc^2 divided by Boltzmann's constant, which must be reached before particles of this type can be created out of radiation energy.

For instance, the lightest known material particles are the electron e^- and the positron e^+. The positron is the "antiparticle" of the electron—that is, it has opposite electrical charge (positive instead of negative) but the same mass and spin. When a positron collides with an electron, the charges can cancel, with the energy in the two particles' masses appearing as pure radiation. This, of course, is why positrons are so rare in ordinary life—they just don't live very long before finding an electron and annihilating. (Positrons were discovered in cosmic rays in 1932.) The annihilation process can also run backward—two photons with sufficient energy can collide and produce an electron-positron pair, the photon energies being converted into the electron and positron masses.

In order for two photons to produce an electron and a positron in a head-on collision, the energy of each photon must exceed the "rest energy" mc^2 in an electron or a positron mass. This energy is 0.511003 million electron volts. To find the threshold temperature at which photons would have a fair chance of having this much energy, we divide the energy by Boltzmann's constant (0.00008617 electron volts per degree Kelvin) and find a threshold temperature of 6 thousand million degrees Kelvin (6×10^9 ° K). At any higher temperature electrons and positrons would have been freely created in collisions of photons with each other, and would therefore be present in very large numbers.

(Incidentally, the threshold temperature of 6×10^9 ° K that we have deduced for the creation of electrons and positrons out of radiation is much higher than any temperature we normally encounter in the present universe. Even the center of the sun is only at a temperature of about 15 million degrees. This is why we are not used to seeing electrons and positrons popping out of empty space whenever the light is bright.)

Similar remarks apply for every type of particle. It is a fundamental rule of modern physics that for every type of particle in nature there is a corresponding "antiparticle," with precisely the same mass and spin, but with opposite electrical charge. The only exception is for certain purely neutral particles, like the photon itself, which can be thought of as being their own antiparticles. The relation between particle and antiparticle is reciprocal: the positron is the antiparticle of the electron, and the electron is the antiparticle of the positron. Given enough energy, it is always possible to create any kind of particle-antiparticle pair in collisions of pairs of photons.

(The existence of antiparticles is a direct mathematical consequence of the principles of quantum mechanics and Einstein's Special Theory of Relativity. The existence of the antielectron was first deduced theoretically by Paul Adrian Maurice Dirac in 1930. Not wanting to introduce an unknown particle into his theory, he identified the antielectron with the only positively charged particle then known, the proton. The discovery of the positron in 1932 verified the theory of antiparticles, and also showed that the proton is not the antiparticle of the electron; it has its own antiparticle, the antiproton, discovered in the 1950s at Berkeley.)

The next lightest particle types after the electron and positron are the muon, or μ^-, a kind of unstable heavy electron, and *its* antiparticle, the μ^+. Just as for electrons and positrons, the μ^- and μ^+ have opposite electrical charge but equal mass, and can be created in collisions of photons with each other. The μ^- and μ^+ each have a rest energy mc^2 equal to 105.6596 million electron volts, and dividing by Boltzmann's constant, the corresponding threshold temperature is 1.2 million million degrees (1.2×10^{12} ° K). Corresponding threshold temperatures for other particles are given in Table One on page 156. By inspection of this table we can tell which particles could have been present in large numbers at various times in the history of the universe: they are just the particles whose threshold temperatures were below the temperature of the universe at that time.

How many of these material particles actually were present at temperatures above the threshold temperature? Under the conditions of high temperature and density that prevailed in the early universe, the number of particles was governed by the basic condition of thermal equilibrium: the number of particles must have been just high enough so that precisely as many were being destroyed each second as were being created. (That is, demand equals supply.) The rate at which any given particle-antiparticle pair will annihilate into two photons is about equal to the rate at which any given pair of photons of the same energy will turn into such a particle and antiparticle. Hence, the condition of thermal equilibrium requires that the number of particles of each type, whose threshold temperature is below the actual temperature, should be about equal to the number of photons. If there are fewer particles than photons, they will be created faster than they are destroyed, and their number will rise; if there are more particles than photons, they will be destroyed faster than they are created, and their number will drop. For instance, at temperatures above the threshold of 6,000 million degrees the number of electrons and positrons must have been about the same as the number of photons, and the universe at these times can be considered to be composed predominantly of photons, electrons, and positrons, not just photons alone.

However, at temperatures above the threshold temperature, a material particle behaves much like a photon. Its average energy is roughly equal to the temperature times Boltzmann's constant, so that high above the threshold temperature its average energy is much larger than the energy in the particle's mass, and the mass can be neglected. Under such conditions the pressure and energy density contributed by material particles of a given type are simply proportional to the fourth power of the temperature, just as for photons. Thus, we can think of the universe at any given time as being composed of a variety of types of "radiation," one type for each species of particle whose threshold temperature was below the cosmic temperature at that time. In particular, the energy density of the universe at any time is proportional to the fourth power of the temperature *and* to the number of species of particles whose threshold temperature is below the cosmic temperature at that time. Conditions of this sort, with temperatures so high that particle-antiparticle pairs are as common in thermal equilibrium as photons, do not exist anywhere in the present universe, except perhaps in the cores of exploding stars. However, we have enough confidence in our knowledge of statistical mechanics to feel safe in making theories about what must have happened under such exotic conditions in the early universe.

To be precise, we should keep in mind that an antiparticle like the positron (e^+) counts as a distinct species. Also, particles like photons and electrons exist in two distinct states of spin, which should be counted as separate species. Finally, particles like the electron (but not the photon) obey a special rule, the "Pauli exclusion principle," which prohibits two particles from occupying the same state; this rule effectively lowers their contribution to the total energy density by a factor of seven-eighths. (It is the exclusion principle that prevents all the electrons in an atom from falling into the same lowest-energy shell; it is therefore responsible for the complicated shell structure of atoms revealed in the periodic table of the elements.) The effective number of species for each type of particle is listed along with the threshold temperatures in Table One on page 156. The energy density of the universe at a given temperature is proportional to the fourth power of the temperature and to the *effective* number of species of particles whose threshold temperatures lie below the temperature of the universe.

Now let's ask *when* the universe was at these elevated temperatures. It is the balance between the gravitational field and the outward momentum of the contents of the universe that governs the rate of expansion of the universe. And it is the total energy density of photons, electrons, positrons, etc., that provided the source of the gravitational field of the universe at early times. We have seen that the energy density of the universe depended essentially only on the temperature, so the cosmic temperature can be used as a sort of clock, cooling instead of ticking as the universe expands. To be more specific, it can be shown that the time required for the energy density of the universe to fall from one value to another is proportional to the difference of the reciprocals of the square roots of the energy densities. (See mathematical note 3, p. 170.) But we have seen that the energy density is proportional to the fourth power of the temperature, and to the number of species of particles with threshold temperatures below the actual temperature. Hence, as long as the temperature does not cross any "threshold" values, *the time that it takes for the universe to cool from one temperature to another is proportional to the difference of the inverse squares of these temperatures.* For instance, if we start at a temperature of 100 million degrees (well below the threshold temperature for electrons) and find that it took 0.06 years (or 22 days) for the temperature to drop to 10 million degrees, then it took another six years for the temperature to drop to one million degrees, another 600 years for the temperature to drop to 100,000 degrees, and so on. The whole time that it took the universe to cool from 100 million degrees to 3,000° K (i.e., to the point where the contents of the universe were just about to become transparent to radiation) was 700,000 years. (See figure 8.) Of course, when I write here of "years" I mean a certain number of absolute time units, as, for instance, a certain number of periods in which an electron makes an orbit around the nucleus in a hydrogen

86

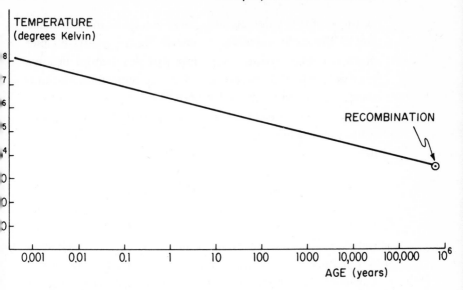

Figure 8. *The Radiation-Dominated Era*. The temperature of the universe is shown as a function of time, for the period from just after the end of nucleosynthesis to the recombination of nuclei and electrons into atoms.

atom. We are dealing with an era long before the earth would begin its tours around the sun.

If the universe in the first few minutes was really composed of precisely equal numbers of particles and antiparticles, they would all have annihilated as the temperature dropped below 1,000 million degrees, and nothing would be left but radiation. There is very good evidence against this possibility—we are here! There must have been some excess of electrons over positrons, of protons over antiprotons, and of neutrons over antineutrons, in order that there would be something left over after the annihilation of particles and antiparticles to furnish the matter of the present universe. Up to this point in this chapter I have purposely ignored the comparatively small amount of this leftover matter. This is a good approximation if all we want is to calculate the energy density or the expansion rate of the early universe; we saw in the preceding chapter that the energy density of nuclear particles did not become comparable to the energy density of radiation until the universe had cooled to about 4,000° K. However, the small seasoning of leftover electrons and nuclear particles has a special claim to our attention, because they dominate the contents of the present universe, and in particular, because they are the main constituents of the author and the reader.

As soon as we admit the possibility of an excess of matter over antimatter in the first few minutes, we open up the problem of determining a detailed list of ingredients for the early universe. There are literally hundreds of so-called elementary particles on the list published every six months by the Lawrence Berkeley Laboratory. Are we going to have to specify the amounts of each one of these types of particle? And why stop at elementary particles—do we also have to specify the numbers of different types of atoms, of molecules, of salt and pepper? In this case, we might well decide that the universe is too complicated and too arbitrary to be worth understanding.

Fortunately, the universe is not that complicated. In order to see how it is possible to write a recipe for its contents, it is necessary to think a little more about what is meant by the condition of thermal equilibrium. I have already emphasized how important it is that the universe has passed through a state of thermal equilibrium—it is what allows us to speak with such confidence about the contents of the universe at any given time. Our discussion so far in this chapter has amounted to a series of applications of the known properties of matter and radiation in thermal equilibrium.

When collisions or other processes bring a physical system to a state of thermal equilibrium, there are always some quantities whose values do not change. One of these "conserved quantities" is the total energy; even though collisions may transfer energy from one particle to another, they never change the total energy of the particles participating in the collision. For each such conservation law there is a quantity that must be specified before we can work out the properties of a system in thermal equilibrium—obviously, if some quantity does not change as a system approaches thermal equilibrium, its value cannot be deduced from the conditions for equilibrium, but must be specified in advance. The really remarkable thing about a system in thermal equilibrium is that *all* its properties are uniquely determined once we specify the values of the conserved quantities. The universe has passed through a state of thermal equilibrium, so to give a complete recipe for the contents of the universe at early times, all we need is to know what were the physical quantities which were conserved as the universe expanded, and what were the values of these quantities.

Usually, as a substitute for specifying the total energy content of a system in thermal equilibrium, we specify the temperature. For the kind of system we have mostly been considering up till now, consisting solely of radiation and equal numbers of particles and antiparticles, the temperature is all that need be given in order to work out the equilibrium properties of the system. But in general there are other conserved quantities in addition to the energy, and it is necessary to specify the densities of each one.

For instance, in a glass of water at room temperature, there are continual reactions in which a water molecule breaks up into a hydrogen ion (a bare proton, the nucleus of hydrogen with the electron stripped off) and a hydroxyl ion (an oxygen atom bound to a hydrogen atom, with an extra electron), or in which hydrogen and hydroxyl ions rejoin to form water molecules. Note that in each such reaction the disappearance of a water molecule is accompanied by the appearance of a hydrogen ion, and vice versa, while hydrogen ions and hydroxyl ions always appear or disappear together. Thus, the conserved quantities are the total number of water molecules *plus* the number of hydrogen ions, and the number of hydrogen ions *minus* the number of hydroxyl ions. (Of course, there are other conserved quantities, like the total number of water molecules plus hydroxyl ions, but these are just simple combinations of the two fundamental conserved quantities.) The properties of our glass of water can be completely determined if we specify that the temperature is 300° K (room temperature on the Kelvin scale), that the density of water molecules plus hydrogen ions is 3.3×10^{22} molecules or ions per cubic centimeter (roughly corresponding to water at sea level pressures), and that the density of hydrogen ions *minus* hydroxyl ions is zero (corresponding to zero net charge). For instance, it turns out that under these conditions there is one hydrogen ion for about every ten million (10^7) water molecules—this is what is meant by the statement that the pH of water is 7. Note that we do not have to specify this in our recipe for a glass of water; we deduce the proportion of hydrogen ions from the rules for thermal equilibrium. On the other hand, we cannot deduce the densities of the conserved quantities from the conditions for thermal equilibrium—for instance, we can make the density of water molecules plus hydrogen ions a little greater or less than 3.3×10^{22} molecules per cubic centimeter by raising or lowering the pressure—so we need to specify them in order to know what is in our glass.

This example also helps us to understand the shifting meaning of what we call "conserved" quantities. For instance, if our water is at a temperature of millions of degrees, as inside a star, then it is very easy for molecules or ions to dissociate, and for the constituent atoms to lose their electrons. The conserved quantities are then the numbers of electrons and of oxygen and hydrogen nuclei. The density of water molecules plus hydroxyl atoms under these conditions has to be *calculated* from the rules of statistical mechanics rather than specified in advance; of course, it turns out to be quite small. (Snowballs are rare in hell.) Actually, nuclear reactions do occur under these conditions, so even the numbers of nuclei of each species are not absolutely fixed, but these numbers change so slowly that a star can be regarded as evolving gradually from one equilibrium state to another.

Ultimately, at the temperatures of several thousand million degrees that we encounter in the early universe, even atomic nuclei dissociate readily into their constituents, protons and neutrons. Reactions occur so rapidly that matter and antimatter can easily be created out of pure energy, or annihilated back again. Under these conditions the conserved quantities are not the numbers of particles of any specific kind. Instead, the relevant conservation laws are reduced to just that small number which (as far as we know) are respected under all possible conditions. There are believed to be just three conserved quantities whose densities must be specified in our recipe for the early universe:

1. *Electric Charge.* We can create or destroy pairs of particles with equal and opposite electric charge, but the *net* electric charge never changes. (We can be more certain about this conservation law than about any of the others, because if charge were not conserved, the accepted Maxwell theory of electricity and magnetism would make no sense.)
2. *Baryon Number.* "Baryon" is an inclusive term which in-

cludes the nuclear particles, protons and neutrons, together with somewhat heavier unstable particles known as hyperons. Baryons and antibaryons can be created or destroyed in pairs; and baryons can decay into other baryons, as in the "beta decay" of a radioactive nucleus in which a neutron changes into a proton, or vice versa. However, the total number of baryons *minus* the number of antibaryons (antiprotons, antineutrons, antihyperons) never changes. We therefore attribute a "baryon number" of $+1$ to the proton, neutron, and hyperons, and a "baryon number" of -1 to the corresponding antiparticles; the rule is then that the total baryon number never changes. Baryon number does not seem to have any dynamical significance like charge; as far as we know there is nothing like an electric or magnetic field produced by baryon number. Baryon number is a bookkeeping device—its significance lies wholly in the fact that it is conserved.

3. *Lepton Number.* The "leptons" are the light negatively charged particles, the electron and muon, plus an electrically neutral particle of zero mass called the neutrino, and their antiparticles, the positron, antimuon, and antineutrino. Despite their zero mass and charge, neutrinos and antineutrinos are no more fictitious than photons; they carry energy and momentum like any other particle. Lepton number conservation is another bookkeeping rule—the total number of leptons minus the total number of antileptons never changes. (In 1962 experiments with beams of neutrinos revealed that there are really at least two types of neutrino, an "electron type" and a "muon type," and two types of lepton number: electron lepton number is the total number of electrons plus electron-type neutrinos, minus the number of their antiparticles, while muon lepton number is the total number of muons plus muon-type neutrinos, minus the number of their antipar-

ticles. Both seem to be absolutely conserved, but this is not know with great certainty.)

A good example of the working of these rules is furnished by the radioactive decay of a neutron n into a proton p, an electron e^-, and an (electron-type) antineutrino $\bar{\nu}_e$. The values of the charge, baryon number, and lepton number of each particle are as follows:

	n	$\rightarrow$	p	$+$	e^-	$+$	$\bar{\nu}_e$
Charge	0		1		-1		0
Baryon Number	$+1$		$+1$		0		0
Lepton Number	0		0		$+1$		-1

The reader can easily check that the sum of the values of any conserved quantity for the particles in the final state equals the value for the same quantity in the initial neutron. This is what we mean by these quantities being conserved. The conservation laws are far from empty, for they tell us that a great many reactions do *not* occur, such as the forbidden decay process in which a neutron decays into a proton, an electron, and more than one antineutrino.

To complete our recipe for the contents of the universe at any given time, we must thus specify the charge, baryon number, and lepton number per unit volume as well as the temperature at that time. The conservation laws tell us that in any volume which expands with the universe the values of these quantities remain fixed. Thus, the charge, baryon number, and lepton number *per unit volume* simply vary with the inverse cube of the size of the universe. But the number of photons per unit volume also varies with the inverse cube of the size of the universe. (We saw in Chapter III that the number of photons per unit volume is proportional to the cube of the temperature, while, as remarked at the beginning of this chapter, the temperature varies with the inverse size of the universe.) Therefore, the charge, baryon number, and lepton number *per photon* remain fixed, and our recipe can be given once and for all by specifying the values of the conserved quantities as a ratio to the number of photons.

(Strictly speaking, the quantity which varies as the inverse cube of the size of the universe is not the number of photons per unit volume but the *entropy* per unit volume. Entropy is a fundamental quantity of statistical mechanics, related to the degree of disorder of a physical system. Aside from a conventional numerical factor, the entropy is given to a good enough approximation by the total number of all particles in thermal equilibrium, material particles as well as photons, with different species of particles given the weights shown in Table One on page 156. The constants that we really should use to characterize our universe are the ratios of charge to entropy, baryon number to entropy, and lepton number to entropy. However, even at very high temperatures the number of material particles is at most of the same order of magnitude as the number of photons, so we will not be making a serious error if we use the number of photons instead of the entropy as our standard of comparison.)

It is easy to estimate the cosmic charge per photon. As far as we know, the average density of electric charge is zero throughout the universe. If the earth and the sun had an excess of positive over negative charges (or vice versa) of only one part in a million million million million million million (10^{36}), the electrical repulsion between them would be greater than their gravitational attraction. If the universe is finite and closed, we can even promote this observation to the status of a theorem: The net charge of the universe must be zero, for otherwise the lines of electrical force would wind round and round the universe, building up to an infinite electric field. But whether the universe is open or closed, it is safe to say that the cosmic electric charge per photon is negligible.

The baryon number per photon is also easy to estimate. The only *stable* baryons are the nuclear particles, the proton and neutron, and their antiparticles, the antiproton and antineutron. (The free neutron is actually unstable, with an average life of 15.3 minutes, but nuclear forces make the neutron absolutely stable in the atomic nuclei of ordinary matter.) Also, as far as we know, there is no appreciable amount of antimatter in the universe. (More about this later.) Hence, the baryon number of any part of the present universe is essentially equal to the number of nuclear particles. We observed in the preceding chapter that there is now one nuclear particle for every 1,000 million photons in the microwave radiation background (the exact figure is uncertain), so the baryon number per photon is about one thousand-millionth (10^{-9}).

This is really a remarkable conclusion. To see its implications, consider a time in the past when the temperature was above ten million million degrees (10^{13} ° K), the threshold temperature for neutrons and protons. At that time the universe would have contained plenty of nuclear particles and antiparticles, about as many as photons. But the baryon number is the *difference* between the numbers of nuclear particles and antiparticles. If this difference were 1,000 million times smaller than the number of photons, and hence also about 1,000 million times smaller than the *total* number of nuclear particles, then the number of nuclear particles would have exceeded the number of antiparticles by only one part in 1,000 million. In this view, when the universe cooled below the threshold temperature for nuclear particles, the antiparticles all annihilated with corresponding particles, leaving the tiny excess of particles over antiparticles as a residue which would eventually turn into the world we know.

The occurrence in cosmology of a pure number as small as one part per 1,000 million has led some theorists to suppose that the number really is zero—that is, that the universe really contains an equal amount of matter and antimatter. Then the fact that the baryon number per photon appears to be one part in 1,000 million would have to be explained by supposing that, at some time before the cosmic temperature dropped below the threshold temperature for nuclear particles, there was a segregation of the universe into different domains, some with a slight excess (a few parts per 1,000 million) of matter over antimatter, and others with a slight excess of antimatter over matter. After the temperature dropped and as many particle-antiparticle pairs as possible annihilated, we would be left with a universe consisting of domains of pure matter and domains of pure antimatter. The trouble with this idea is that no one has seen signs of appreciable amounts of antimatter anywhere in the universe. The cosmic rays that enter our earth's upper atmosphere are believed to come in part from great distances in our galaxy, and perhaps in part from outside our galaxy as well. The cosmic rays are overwhelmingly matter rather than antimatter—in fact, no one has yet observed an antiproton or an antinucleus in the cosmic rays. In addition, we do not observe the photons that would be produced from annihilation of matter and antimatter on a cosmic scale.

Another possibility is that the density of photons (or, more properly, of entropy) has not remained proportional to the inverse cube of the size of the universe. This could happen if there were some sort of departure from thermal equilibrium, some sort of friction or viscosity which could have heated the universe and produced extra photons. In this case, the baryon number per photon might have started at some reasonable value, perhaps around one, and then dropped to its present low value as more photons were produced. The trouble is that no one has been able to suggest any detailed mechanism for producing these extra photons. I tried to find one some years ago, with utter lack of success.

In what follows I will ignore all these "nonstandard" possibilities, and will simply assume that the baryon number per photon is what it seems to be: about one part in 1,000 million.

What about the lepton number density of the universe? The fact that the universe has no electric charge tells us immediately that there is now precisely one negatively charged electron for each positively charged proton. About 87 percent of the nuclear particles in the present universe are protons, so the number of electrons is close to the total number of nuclear particles. If electrons were the only leptons in the present universe, we could conclude immediately that the lepton number per photon is roughly the same as the baryon number per photon.

However, there is one other kind of stable particle besides the electron and positron that carries a nonzero lepton number. The neutrino and its antiparticle the antineutrino are electrically neutral massless particles, like the photon, but with lepton numbers $+1$ and -1, respectively. Thus, in order to determine the lepton number density of the present universe, we have to know something about the populations of neutrinos and antineutrinos.

Unfortunately this information is extraordinarily difficult to come by. The neutrino is like the electron in that it does not feel the strong nuclear force that keeps protons and neutrons inside the atomic nucleus. (I will sometimes use "neutrino" to mean neutrino or antineutrino.) However, unlike the electron, it is electrically neutral, so it also does not feel electrical or magnetic forces like those which keep electrons inside the atom. In fact, neutrinos do not respond much to any sort of force at all. They do respond like everything else in the universe to the force of gravitation, and they also feel the weak force responsible for radioactive processes like the neutron decay mentioned earlier (see p. 93), but these forces produce only a tiny interaction with ordinary matter. The example that is usually quoted to show how weakly neutrinos interact is that, in order to have an appreciable chance of stopping or scattering any given neutrino produced in some radioactive process, we would need to place several light years of lead in its path. The sun is continually radiating neutrinos, produced when protons turn into neutrons in the nuclear reactions in the sun's core; these neutrinos shine down on us during the day and shine *up* on us at night, when the sun is on the other side of the earth, because the earth is utterly transparent to them. Neutrinos were hypothesized by Wolfgang Pauli long before they were observed, as a means of accounting for the energy balance in a process like neutron decay. It is only since the late 1950s that it has been possible to detect neutrinos or antineutrinos directly, by producing such vast quantities in nuclear reactors or particle accelerators that a few hundred actually stop within the detecting apparatus.

Given this extraordinary weakness of interaction, it is easy to understand that enormous numbers of neutrinos and antineutrinos could be filling the universe around us without our having any hint of their presence. It is possible to set some vague upper limits on the number of neutrinos and antineutrinos: if these particles were too numerous, then certain weak nuclear decay processes would be slightly affected, and in addition the cosmic expansion would be decelerating more rapidly than is observed. However, these upper limits do not rule out the possibility that there are about as many neutrinos and/or antineutrinos as photons, and with similar energies.

Despite these remarks, it is usual for cosmologists to assume that the lepton number (the numbers of electrons, muons, and neutrinos, *minus* the numbers of their corresponding antiparticles) per photon is small, much less than one. This is purely on the basis of analogy—the baryon number per photon is small, so why should the lepton number per photon not also be small? This is one of the least certain of the assumptions that go into the "standard model," but fortunately, even if it were false, the general picture we derive would be changed only in detail.

Of course, above the threshold temperature for electrons there were lots of leptons and antileptons—about as many electrons and positrons as photons. Also, under these conditions, the universe was so hot and dense that even the ghostly neutrinos reached thermal equilibrium, so that there were also about as many neutrinos and antineutrinos as photons. The assumption made in the standard model is that the lepton number, the *difference* in the number of leptons and antileptons, is and was much smaller than the number of photons. There may have been some small excess of leptons over antileptons, like the small excess of baryons over antibaryons mentioned earlier, which has survived to the present time. In addition, the neutrinos and antineutrinos interact so weakly that large numbers of them may have escaped annihilation, in which case there would now be nearly equal numbers of neutrinos and antineutrinos, comparable to the number of photons. We will see in the next chapter that this is indeed believed to be the case, but there does not seem to be the slightest chance in the foreseeable future of observing the vast number of neutrinos and antineutrinos around us.

This then in brief is our recipe for the contents of the early universe. Take a charge per photon equal to zero, a baryon number per photon equal to one part in 1,000 million, and a lepton number per photon uncertain but small. Take the temperature at any given time to be greater than the temperature 3° K of the present radiation background by the ratio of the present size of the universe to the size at that time. Stir well, so that the detailed distributions of particles of various types are determined by the requirements of thermal equilibrium. Place in an expanding universe, with a rate of expansion governed by the gravitational field produced by this medium. After a long enough wait, this concoction should turn into our present universe.

V

THE FIRST
THREE MINUTES

WE ARE NOW prepared to follow the course of cosmic evolution through its first three minutes. Events move much more swiftly at first than later, so it would not be useful to show pictures spaced at equal time intervals, like an ordinary movie. Instead, I will adjust the speed of our film to the falling temperature of the universe, stopping the camera to take a picture each time that the temperature drops by a factor of about three.

Unfortunately, I cannot start the film at zero time and infinite temperature. Above a threshold temperature of fifteen hundred thousand million degrees Kelvin (1.5×10^{12} ° K), the universe would contain large numbers of the particles known as pi mesons, which weigh about one-seventh as much as a nuclear particle. (See Table One on p. 156.) Unlike the electrons, positrons, muons, and neutrinos, the pi mesons interact very strongly with each other and with nuclear particles—in fact, the continual exchange of pi mesons among nuclear particles is responsible for most of the attractive force which holds atomic nuclei together. The presence of large numbers of such strongly interacting particles makes it extraordinarily difficult to calculate the behavior of matter at super-high temperatures, so to avoid such difficult mathematical problems I will start the story in this chapter at about one-hundredth of a second after the beginning, when the temperature had cooled to a mere hundred thousand million degrees Kelvin, safely below the threshold temperatures for pi mesons, muons, and all heavier particles. In Chapter VII I will say a little about what theoretical physicists think may have been going on closer to the very beginning.

With these understandings, let us now start our film.

FIRST FRAME. The temperature of the universe is 100,000 million degrees Kelvin (10^{11} ° K). The universe is simpler and easier to describe than it ever will be again. It is filled with an undifferentiated soup of matter and radiation, each particle of which collides very rapidly with the other particles. Thus, despite its rapid expansion, the universe is in a state of nearly perfect thermal equilibrium. The contents of the universe are therefore dictated by the rules of statistical mechanics, and do not depend at all on what went before the first frame. All we need to know is that the temperature is 10^{11} ° K, and that the conserved quantities—charge, baryon number, lepton number—are all very small or zero.

The abundant particles are those whose threshold temperatures are below 10^{11} ° K; these are the electron and its anti-particle, the positron, and of course the massless particles, the photon, neutrinos, and antineutrinos. (Again, see Table One on p. 156.) The universe is so dense that even the neutrinos, which can travel for years through lead bricks without being scattered, are kept in thermal equilibrium with the electrons, positrons, and photons by rapid collisions with them and with each other. (Again, I will sometimes simply refer to "neutrinos" when I mean neutrinos and antineutrinos.)

Another great simplification—the temperature of 10^{11} ° K is far above the threshold temperature for electrons and positrons. It follows that these particles, as well as the photons and neutrinos, are behaving just like so many different kinds of radiation. What is the energy density of these various kinds of radiation? According to Table One on page 156, the electrons and positrons together contribute 7/4 as much energy as the photons, and the neutrinos and antineutrinos contribute the same as the electrons and positrons, so the total energy density is greater than the energy density for pure electromagnetic radiation at this temperature, by a factor

$$\frac{7}{4} + \frac{7}{4} + 1 = \frac{9}{2}$$

The Stefan-Boltzmann law (see Chapter III) gives the energy density of electromagnetic radiation at a temperature of 10^{11} ° K as 4.72×10^{44} electron volts per liter, so the total energy density of the universe at this temperature was 9/2 as great, or 21×10^{44} electron volts per liter. This is equivalent to a mass density of 3.8 thousand million kilograms per liter, or 3.8 thousand million times the density of water under normal terrestrial conditions. (When I speak of a given energy as being equivalent to a given mass, I mean of course that this is the energy that would be released according to the Einstein formula $E = mc^2$, if the mass were converted entirely to energy.) If Mt. Everest were made of matter this dense, its gravitational attraction would destroy the earth.

The universe at the first frame is rapidly expanding and cooling. Its rate of expansion is set by the condition that every bit of the universe is traveling just at escape velocity away from any arbitrary center. At the enormous density of the first frame, the escape velocity is correspondingly high—the characteristic time for expansion of the universe is about 0.02 seconds. (See mathematical note 3, p. 169. The "characteristic expansion time" can be roughly defined as 100 times the length of time in which the size of the universe would increase 1 percent. To be more precise, the characteristic expansion time at any epoch is the reciprocal of the Hubble "constant" at that epoch. As remarked in Chapter II, the age of the universe is always less than the characteristic expansion time, because gravitation is continually slowing down the expansion.)

There are a small number of nuclear particles at the time of the first frame, about one proton or neutron for every 1,000 million photons or electrons or neutrinos. In order eventually to predict the abundances of the chemical elements formed in the early universe, we will also need to know the relative proportions of protons and neutrons. The neutron is heavier than the proton, with a mass difference between them equivalent to an energy of 1.293 million electron volts. However, the characteristic energy of the electrons, positrons, and so on, at a temperature of 10^{11} ° K is much larger, about 10 million electron volts (Boltzmann's constant times the temperature). Thus, collisions of neutrons or protons with the much more numerous electrons, positrons, and so on, will produce rapid transitions of protons to neutrons and vice versa. The most important reactions are

Antineutrino plus proton yields positron plus neutron
(and vice versa)

Neutrino plus neutron yields electron plus proton
(and vice versa)

Under our assumption that the net lepton number and charge per photon are very small, there are almost exactly as many neutrinos as antineutrinos, and as many positrons as electrons, so that the transitions from proton to neutron are just as fast as the transitions from neutron to proton. (The radioactive decay of the neutron can be ignored here because it takes about 15 minutes, and we are working now on a time scale of hundredths of seconds.) Equilibrium thus requires that the numbers of protons and neutrons be just about equal at the first frame. These nuclear particles are not yet bound into nuclei; the energy required to break up a typical nucleus altogether is only six to eight million electron volts per nuclear particle; this is less than the characteristic thermal energies at 10^{11} ° K, so complex nuclei are destroyed as fast as they form.

It is natural to ask how large the universe was at very early times. Unfortunately we do not know, and we are not even sure that this question has any meaning. As indicated in Chapter II, the universe may well be infinite now, in which case it was also infinite at the time of the first frame, and will always be infinite. On the other hand, it is possible that the universe now has a finite circumference, sometimes estimated to be about 125 thousand million light years. (The circumference is the distance one must travel in a straight line before finding oneself back where one started. This estimate is based

on the present value of the Hubble constant, under the sup-
position that the density of the universe is about twice its "critical" value.) Since the temperature of the universe falls in inverse proportion to its size, the circumference of the universe
at the time of the first frame was less than at present by the
ratio of the temperature then (10^{11} ° K) to the present temperature (3° K); this gives a first-frame circumference of about four
light years. None of the details of the story of cosmic evolution
in the first few minutes will depend on whether the circumference of the universe was infinite or only a few light years.

SECOND FRAME. The temperature of the universe is 30,000
million degrees Kelvin (3×10^{10} ° K). Since the first frame,
0.11 seconds have elapsed. Nothing has changed qualitatively—the contents of the universe are still dominated by
electrons, positrons, neutrinos, antineutrinos, and photons,
all in thermal equilibrium, and all high above their threshold
temperatures. Hence the energy density has dropped simply
like the fourth power of the temperature, to about 30 million
times the energy density contained in the rest mass of ordinary
water. The rate of expansion has dropped like the square of the
temperature, so that the characteristic expansion time of the
universe has now lengthened to about 0.2 seconds. The small
number of nuclear particles is still not bound into nuclei, but
with falling temperature it is now significantly easier for the
heavier neutrons to turn into the lighter protons than vice
versa. The nuclear particle balance has consequently shifted to
38 percent neutrons and 62 percent protons.

THIRD FRAME. The temperature of the universe is 10,000 million degrees Kelvin (10^{10} ° K). Since the first frame, 1.09 seconds have elapsed. About this time the decreasing density and
temperature have increased the mean free time of neutrinos

and antineutrinos so much that they are beginning to behave like free particles, no longer in thermal equilibrium with the electrons, positrons, or photons. From now on they will cease to play any active role in our story, except that their energy will continue to provide part of the source of the gravitational field of the universe. Nothing much changes when the neutrinos go out of thermal equilibrium. (Before this "decoupling," the typical neutrino wavelengths were inversely proportional to the temperature, and since the temperature was falling off in inverse proportion to the size of the universe, the neutrino wavelengths were increasing in direct proportion to the size of the universe. After neutrino decoupling the neutrinos will expand freely, but the general red shift will still stretch out their wavelengths in direct proportion to the size of the universe. This shows, incidentally, that it is not too important to determine the precise instant of neutrino decoupling, which is just as well, because it depends on details of the theory of neutrino interactions which are not entirely settled.)

The total energy density is less than it was in the last frame by the fourth power of the ratio of temperatures, so it is now equivalent to a mass density 380,000 times that of water. The characteristic time for expansion of the universe has correspondingly increased to about two seconds. The temperature is now only twice the threshold temperature of electrons and positrons, so they are just beginning to annihilate more rapidly than they can be recreated out of radiation.

It is still too hot for neutrons and protons to be bound into atomic nuclei for any appreciable time. The decreasing temperature has now allowed the proton-neutron balance to shift to 24 percent neutrons and 76 percent protons.

FOURTH FRAME. The temperature of the universe is now 3,000 million degrees Kelvin (3×10^9 ° K). Since the first frame,

13.82 seconds have elapsed. We are now below the threshold temperature for electrons and positrons, so they are beginning rapidly to disappear as major constituents of the universe. The energy released in their annihilation has slowed down the rate at which the universe cools, so that the neutrinos, which do not get any of this extra heat, are now 8 percent cooler than the electrons, positrons, and photons. From now on, when we refer to the temperature of the universe we will mean the temperature of the *photons*. With electrons and positrons rapidly disappearing, the energy density of the universe is now somewhat less than it would be if it were simply falling off like the fourth power of the temperature.

It is now cool enough for various stable nuclei like helium (He⁴) to form, but this does not happen immediately. The reason is that the universe is still expanding so rapidly that nuclei can only be formed in a series of fast two-particle reactions. For instance, a proton and a neutron can form a nucleus of heavy hydrogen, or deuterium, with the extra energy and momentum being carried away by a photon. The deuterium nucleus can then collide with a proton or a neutron, forming either a nucleus of the light isotope, helium three (He³), consisting of two protons and a neutron, or the heaviest isotope of hydrogen, called tritium (H³), consisting of a proton and two neutrons. Finally, the helium three can collide with a neutron, and the tritium can collide with a proton, in both cases forming a nucleus of ordinary helium (He⁴), consisting of two protons and two neutrons. But in order for this chain of reactions to occur, it is necessary to start with the first step, the production of deuterium.

Now, ordinary helium is a tightly bound nucleus, so, as I said, it can indeed hold together at the temperature of the third frame. However, tritium and helium three are much less tightly bound, and deuterium is especially loosely bound. (It takes only a ninth as much energy to pull a deuterium nucleus apart as to pull a single nuclear particle out of a helium nucleus.) At the fourth-frame temperature of 10^{10} ° K, nuclei of deuterium are blasted apart as soon as they form, so heavier nuclei do not get a chance to be produced. Neutrons are still being converted into protons, although much more slowly than before; the balance now is 17 percent neutrons and 83 percent protons.

FIFTH FRAME. The temperature of the universe is now 1,000 million degrees Kelvin (10^9 ° K), only about 70 times hotter than the center of the sun. Since the first frame, three minutes and two seconds have elapsed. The electrons and positrons have mostly disappeared, and the chief constituents of the universe are now photons, neutrinos, and antineutrinos. The energy released in electron-positron annihilation has given the photons a temperature 35 percent higher than that of the neutrinos.

The universe is now cool enough for tritium and helium three as well as ordinary helium nuclei to hold together, but the "deuterium bottleneck" is still at work: nuclei of deuterium do not hold together long enough to allow appreciable numbers of heavier nuclei to be built up. The collisions of neutrons and protons with electrons, neutrinos, and their antiparticles have now pretty well ceased, but the decay of the free neutron is beginning to be important; in each 100 seconds, 10 percent of the remaining neutrons will decay into protons. The neutron-proton balance is now 14 percent neutrons, 86 percent protons.

A LITTLE LATER. At some time shortly after the fifth frame, a dramatic event occurs: the temperature drops to the point at which deuterium nuclei can hold together. Once the deu-

E

terium bottleneck is passed, heavier nuclei can be built up very rapidly by the chain of two-particle reactions described in the fourth frame. However, nuclei heavier than helium are not formed in appreciable numbers because of other bottle-necks: there are no stable nuclei with five or eight nuclear par-ticles. Hence, as soon as the temperature reaches the point where deuterium can form, almost all of the remaining neu-trons are immediately cooked into helium nuclei. The precise temperature at which this happens depends slightly on the number of nuclear particles per photon, because a high par-ticle density would make it a little easier for nuclei to form. (This is why I had to identify this moment imprecisely as "a little later" than the fifth frame.) For 1,000 million photons per nuclear particle, nucleosynthesis will begin at a tempera-ture of 900 million degrees Kelvin (0.9×10^9 ° K). At this time, three minutes forty-six seconds have passed since the first frame. (The reader will have to forgive my inaccuracy in calling this book *The First Three Minutes*. It sounded better than *The First Three and Three-quarter Minutes*.) Neutron decay would have shifted the neutron-proton balance just be-fore nucleosynthesis began to 13 percent neutrons, 87 percent protons. After nucleosynthesis, the fraction by weight of helium is just equal to the fraction of all nuclear particles that are bound into helium; half of these are neutrons, and essen-tially all neutrons are bound into helium, so the fraction by weight of helium is simply twice the fraction of neutrons among nuclear particles, or about 26 percent. If the density of nuclear particles is a little higher, nucleosynthesis begins a little earlier, when not so many neutrons would have decayed, so slightly more helium is produced, but probably not more than 28 percent by weight. (See figure 9.)

We have now reached and exceeded our planned running

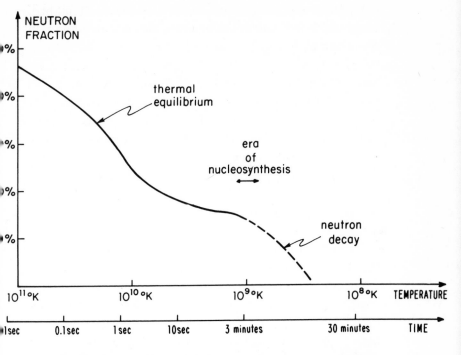

Figure 9. *The Shifting Neutron-Proton Balance.* The fraction of neutrons to all nuclear particles is shown as a function both of temperature and of time. The part of the
curve marked "thermal equilibrium" describes the period in which densities and temperature are so high that thermal equilibrium is maintained among all particles; the
neutron fraction here can be calculated from the neutron-proton mass difference,
using the rules of statistical mechanics. The part of the curve marked "neutron decay"
describes the period in which all neutron-proton conversion processes have ceased,
except for the radioactive decay of the free neutron. The intervening part of the curve
depends on detailed calculations of weak-interaction transition rates. The dashed part
of the curve shows what would happen if nuclei were somehow prevented from forming. Actually, at a time somewhere within the period indicated by the arrow marked
"era of nucleosynthesis," neutrons are rapidly assembled into helium nuclei, and the
neutron-proton ratio is frozen at the value it has at that time. This curve can also be
used to estimate the fraction (by weight) of cosmologically produced helium: for any
given value of the temperature or the time of nucleosynthesis, it is just twice the neutron fraction at that time.

time, but in order to see better what has been accomplished, let us take a last look at the universe after one more drop in temperature.

SIXTH FRAME. The temperature of the universe is now 300 million degrees Kelvin (3×10^8 ° K). Since the first frame, 34 minutes and 40 seconds have elapsed. The electrons and positrons are now completely annihilated except for the small (one part in 1,000 million) excess of electrons needed to balance the charge of the protons. The energy released in this annihilation has now given the photons a temperature permanently 40.1 percent higher than the temperature of the neutrinos. (See mathematical note 6, p. 176.) The energy density of the universe is now equivalent to a mass density 9.9 percent that of water; of this, 31 percent is in the form of neutrinos and antineutrinos and 69 percent is in the form of photons. This energy density gives the universe a characteristic expansion time of about one and one-quarter hours. Nuclear processes have stopped—the nuclear particles are now for the most part either bound into helium nuclei or are free protons (hydrogen nuclei), with about 22 to 28 percent helium by weight. There is one electron for each free or bound proton, but the universe is still much too hot for stable atoms to hold together.

The universe will go on expanding and cooling, but not much of interest will occur for 700,000 years. At that time the temperature will drop to the point where electrons and nuclei can form stable atoms; the lack of free electrons will make the contents of the universe transparent to radiation; and the decoupling of matter and radiation will allow matter to begin to form into galaxies and stars. After another 10,000 million years or so, living beings will begin to reconstruct this story.

This account of the early universe has one consequence that can be immediately tested against observation: the material left over from the first three minutes, out of which the stars must originally have formed, consisted of 22–28 percent helium, with almost all the rest hydrogen. As we have seen, this result depends on the assumption that there is a huge ratio of photons to nuclear particles, which in turn is based on the measured 3° K temperature of the present cosmic microwave radiation background. The first calculation of the cosmological helium production to make use of the measured radiation temperature was carried out by P. J. E. Peebles at Princeton in 1965, shortly after the discovery of the microwave background by Penzias and Wilson. A similar result was obtained independently and at almost the same time in a more elaborate calculation by Robert Wagoner, William Fowler, and Fred Hoyle. This result was a stunning success for the standard model, for there were already at this time independent estimates that the sun and other stars do start their lives as mostly hydrogen, with about 20–30 percent helium!

There is, of course, extremely little helium on earth, but that is just because helium atoms are so light and so chemically inert that most of them escaped the earth ages ago. Estimates of the primordial helium abundance of the universe are based on comparisons of detailed calculations of stellar evolution with statistical analyses of observed stellar properties, plus direct observation of helium lines in the spectra of hot stars and interstellar material. Indeed, as its name indicates, helium was first identified as an element in studies of the spectrum of the sun's atmosphere, carried out in 1868 by J. Norman Lockyer.

During the early 1960s it was noticed by a few astronomers that the abundance of helium in the galaxy is not only large, but does not vary from place to place nearly so much as does the abundance of heavier elements. This of course is just what would be expected if the heavy elements were produced in stars, but helium was produced in the early universe, before any of the stars began to cook. There is still a good deal of uncertainty and variation in estimates of nuclear abundances, but the evidence for a primordial 20–30 percent helium abundance is strong enough to give great encouragement to adherents of the standard model.

In addition to the large amount of helium produced at the end of the first three minutes, there was also a trace of lighter nuclei, chiefly deuterium (hydrogen with one extra neutron) and the light helium isotope He^3, that escaped incorporation into ordinary helium nuclei. (Their abundances were first calculated in the 1967 paper of Wagoner, Fowler, and Hoyle.) Unlike the helium abundance, the deuterium abundance is very sensitive to the density of nuclear particles at the time of nucleosynthesis: for higher densities, nuclear reactions proceeded faster, so that almost all deuterium would have been cooked into helium. To be specific, here are the values of the abundance of deuterium (by weight) produced in the early universe, as given by Wagoner, for three possible values of the ratio of photons to nuclear particles:

Photons/nuclear particle	Deuterium abundance (parts per million)
100 million	0.00008
1,000 million	16
10,000 million	600

Clearly, if we could determine the primordial deuterium abundance that existed before stellar cooking began, we could make a precise determination of the photon-to-nuclear particle ratio; knowing the present radiation temperature of 3°K, we could then determine a precise value for the present nuclear mass density of the universe, and judge whether it is open or closed.

Unfortunately it has been very difficult to determine a truly primordial deuterium abundance. The classic value for the abundance by weight of deuterium in the water on earth is 150 parts per million. (This is the deuterium that will be used to fuel thermonuclear reactors, if thermonuclear reactions can ever be adequately controlled.) However, this is a biased figure; the fact that deuterium atoms are twice as heavy as hydrogen atoms makes it somewhat more likely for them to be bound into molecules of heavy water (HDO), so that a smaller proportion of deuterium than hydrogen would have escaped the earth's gravitational field. On the other hand, spectroscopy indicates a very low abundance of deuterium on the sun's surface—less than four parts per million. This also is a biased figure—deuterium in the outer regions of the sun would have been mostly destroyed by fusing with hydrogen into the light isotope of helium, He^3.

Our knowledge of the cosmic deuterium abundance was put on a much firmer basis by ultraviolet observations in 1973 from the artificial Earth satellite *Copernicus*. Deuterium atoms, like hydrogen atoms, can absorb ultraviolet light at certain distinct wavelengths, corresponding to transitions in which the atom is excited from the state of lowest energy to one of the higher states. These wavelengths depend slightly on the mass of the atomic nucleus, so the ultraviolet spectrum of a star whose light passes to us through an interstellar mixture of hydrogen and deuterium will be crossed with a number of dark absorption lines, each split into two components, one from hydrogen and one from deuterium. The relative darkness of any pair of absorption line components then immediately gives the relative abundance of hydrogen and deuterium in the interstellar cloud. Unfortunately, the earth's atmosphere makes it very difficult to do any sort of ultraviolet astronomy from the ground. The satellite *Copernicus* carried an ultraviolet spectrometer which was used to study absorption lines in the spectrum of the hot star β Centaurus; from their relative intensities, it was found that the interstellar medium between us and β Centaurus contains about 20 parts per million (by weight) of deuterium. More recent observations of ultraviolet absorption lines in the spectra of other hot stars give similar results.

If this 20 parts per million of deuterium was really created in the early universe, then there must have been (and is now) just about 1,100 million photons per nuclear particle (see the table above). At the present cosmic radiation temperature of 3° K there are 550,000 photons per liter, so there must be now about 500 nuclear particles per million liters. This is considerably less than the minimal density for a closed universe, which, as we saw in Chapter II, is about 3,000 nuclear particles per million liters. The conclusion would then be that the universe is open; that is, the galaxies are moving at above escape velocity, and the universe will expand forever. If some of the interstellar medium has been processed in stars which tend to destroy deuterium (as in the sun), then the cosmologically produced deuterium abundance must have been even greater than the 20 parts per million found by the *Copernicus* satellite, so the density of nuclear particles must be even less than 500 particles per million liters, strengthening the conclusion that we live in an open, eternally expanding universe.

I must say that I personally find this line of argument rather unconvincing. Deuterium is not like helium—even though its abundance seems higher than would be expected for a relatively dense closed universe, deuterium is still extremely rare in absolute terms. We can imagine that this much deuterium was produced in "recent" astrophysical phenomena—supernovas, cosmic rays, perhaps even quasi-stellar objects. This is not the case for helium; the 20–30 percent helium abundance could not have been created recently without liberating enormous amounts of radiation that we do not observe. It is argued that the 20 parts per million of deuterium found by *Copernicus* could not have been produced by any conventional astrophysical mechanism without also producing unacceptably large amounts of the other rare light elements: lithium, berylium, and boron. However, I do not see how we are ever going to be sure that this trace of deuterium was not produced by some noncosmological mechanism that no one has thought of yet.

There is one other remnant of the early universe that is present all around us, and yet seems impossible to observe. We saw in the third frame that neutrinos have behaved like free particles since the cosmic temperature dropped below about 10,000 million degrees Kelvin. During this time the neutrino wavelengths have simply expanded in proportion to the size of the universe; the number and energy distribution of the neutrinos have consequently remained the same as they would be in thermal equilibrium, but with a temperature that has dropped in inverse proportion to the size of the universe. This is just about the same as what has happened to photons during this time, even though photons remained in thermal equilibrium far longer than neutrinos. Hence the present neutrino temperature ought to be roughly the same as the present photon temperature. There would therefore be something like 1,000 million neutrinos and antineutrinos for every nuclear particle in the universe.

It is possible to be considerably more precise about this. A little after the universe became transparent to neutrinos, the electrons and positrons began to annihilate, heating the photons but not the neutrinos. In consequence, the present neutrino temperature ought to be a little *less* than the present photon temperature. It is fairly easy to calculate that the neutrino temperature is less than the photon temperature by a factor of the cube root of 4/11, or 71.38 percent; the neutrinos and antineutrinos then contribute 45.42 percent as much energy to the universe as photons. (See mathematical note 6, p. 176.) Although I have not said so explicitly, whenever I have quoted cosmic expansion times previously, I have taken this extra neutrino energy density into account.

The most dramatic possible confirmation of the standard model of the early universe would be the detection of this neutrino background. We have a firm prediction of its temperature; it is 71.38 percent of the photon temperature, or just about 2° K. The only real theoretical uncertainty in the number and energy distribution of neutrinos is in the question of whether the lepton number density is small, as we have been assuming. (Recall that the lepton number is the number of neutrinos and other leptons *minus* the number of antineutrinos and other antileptons.) If the lepton number density is as small as the baryon number density, then the numbers of neutrinos and antineutrinos should be equal to each other, to one part in 1,000 million. On the other hand, if the lepton number density is comparable to the photon number density, then there would be a "degeneracy," an appreciable excess of neutrinos (or antineutrinos) and a deficiency of antineutrinos (or neutrinos). Such a degeneracy would affect the shifting neutron-proton balance in the first three minutes, and hence would change the amounts of helium and deuterium produced cosmologically. Observation of the 2° K cosmic neutrino and antineutrino background would immediately settle the question of whether the universe has a large lepton number, but much more important, it would prove that the standard model of the early universe is really true.

Alas, neutrinos interact so weakly with ordinary matter that no one has been able to devise any method for observing a 2° K cosmic neutrino background. It is a truly tantalizing problem: there are some 1,000 million neutrinos and antineutrinos for every nuclear particle, and yet no one knows how to detect them! Perhaps someday someone will.

In following this account of the first three minutes, the reader may feel that he can detect a note of scientific overconfidence. He might be right. However, I do not believe that scientific progress is always best advanced by keeping an altogether open mind. It is often necessary to forget one's doubts and to follow the consequences of one's assumptions wherever they may lead—the great thing is not to be free of theoretical prejudices, but to have the right theoretical prejudices. And always, the test of any theoretical preconception is in where it leads. The standard model of the early universe has scored some successes, and it provides a coherent theoretical framework for future experimental programs. This does not mean that it is true, but it does mean that it deserves to be taken seriously.

Nevertheless, there *is* one great uncertainty that hangs like a dark cloud over the standard model. Underlying all the calculations described in this chapter is the Cosmological Principle, the assumption that the universe is homogeneous and isotropic. (See p. 21. By "homogeneous" we mean that the universe looks the same to any observer who is carried along by the general expansion of the universe, wherever that observer may be located; by "isotropic" we mean that the universe looks the same in all directions to such an observer.) We know from direct observation that the cosmic microwave radiation background is highly isotropic about us, and from this we infer that the universe has been highly isotropic and homogeneous ever since the radiation went out of equilibrium with matter, at a temperature of about 3,000° K. However, we have no evidence that the Cosmological Principle was valid at earlier times.

It is possible that the universe was initially highly inhomogeneous and anisotropic, but has subsequently been smoothed out by the frictional forces exerted by the parts of the expanding universe on each other. Such a "mixmaster" model has been particularly advocated by Charles Misner of the University of Maryland. It is even possible that the heat generated by the frictional homogenization and isotropization of the universe is responsible for the enormous 1,000 million-to-one present ratio of photons to nuclear particles. However, to the best of my knowledge, no one can say why the universe should have any specific initial degree of inhomogeneity or anisotropy, and no one knows how to calculate the heat produced by its smoothing out.

In my opinion, the appropriate response to such uncertainties is not (as some cosmologists might like) to scrap the standard model, but rather to take it very seriously and to work out its consequences thoroughly, if only in the hope of turning up a contradiction with observation. It is not even clear that a large initial anisotropy and inhomogeneity would have much effect on the story presented in this chapter. It might be that the universe was smoothed out in the first few seconds; in that case the cosmological production of helium and deuterium could be calculated as if the Cosmological Principle were always valid. Even if the anisotropy and inhomogeneity of the universe persisted beyond the era of helium synthesis, the helium and deuterium production in any uniformly expanding clump would depend only on the expansion rate within that clump, and might not be very different from the production calculated in the standard model. It might even be that the whole universe that we can see when we look all the way back to the time of nucleosynthesis is but a homogeneous and isotropic clump within a larger inhomogeneous and anisotropic universe.

The uncertainty surrounding the Cosmological Principle becomes really important when we look back to the very beginning or forward to the final end of the universe. I will continue to rely on this Principle in most of the last two chapters. However, it must always be admitted that our simple cosmological models may only describe a small part of the universe, or a limited portion of its history.

VI

A HISTORICAL DIVERSION

L ET US turn away for a moment from the history of the early universe, and take up the history of the last three decades of cosmological research. I want especially to grapple here with a historical problem that I find both puzzling and fascinating. The detection of the cosmic microwave radiation background in 1965 was one of the most important scientific discoveries of the twentieth century. Why did it have to be made by accident? Or to put it another way, why was there no systematic search for this radiation, years before 1965?

As we saw in the last chapter, the measured present value of the radiation background temperature and mass density of the universe allow us to predict cosmic abundances of the light elements that seem to be in good agreement with observation.

Long before 1965 it would have been possible to run this calculation backward, to predict a cosmic microwave background, and to start to search for it. From the observed present cosmic abundances of about 20–30 percent helium and 70–80 percent hydrogen, it would have been possible to infer that nucleosynthesis must have had to begin at a time when the neutron fraction of nuclear particles had dropped to 10–15 percent. (Recall that the present helium abundance by weight is just twice the neutron fraction at the time of nucleosynthesis.) This value of the neutron fraction was reached when the universe was at a temperature of about 1,000 million degrees Kelvin ($10^{9\circ}$ K). The condition that nucleosynthesis began at this moment would allow one to make a rough estimate of the density of nuclear particles at the temperature of $10^{9\ \circ}$ K, while the density of photons at this temperature can be calculated from the known properties of black-body radiation. Hence the ratio of the numbers of photons and nuclear particles would also be known at this time. But this ratio does not change, so it would also be known equally well at the present time. From observations of the present density of nuclear particles, one could thus predict the present density of photons, and infer the existence of a cosmic microwave radiation background with a present temperature roughly in the range of 1° K to 10° K. If the history of science were so simple and direct as the history of the universe, someone would have made a prediction along these lines in the 1940s or 1950s, and it would have been this prediction that instigated radio astronomers to search for the radiation background. But that is not quite what happened.

In fact, a prediction much along these lines *was* made in 1948, but it did not lead then or later to a search for the radiation. In the late 1940s, a "big bang" cosmological theory was being explored by George Gamow and his colleagues Ralph A. Alpher and Robert Herman. They assumed that the universe started as pure neutrons, and that the neutrons then began to convert to protons through the familiar radioactive decay process in which a neutron spontaneously turns into a proton, an electron, and an antineutrino. At some time in the expansion, it would become cool enough for heavy elements to be built up out of neutrons and protons by a rapid sequence of neutron captures. Alpher and Herman found that to account for the observed present abundances of the light elements, it was necessary to assume a ratio of photons to nuclear particles of the order of 1,000 million. Using estimates of the present cosmic density of nuclear particles, they were then able to predict the existence of a radiation background left over from the early universe, with a present temperature of 5° K!

The original calculations of Alpher, Herman, and Gamow were not correct in all details. As we saw in the preceding chapter, the universe probably started with equal numbers of neutrons and protons, not pure neutrons. Also, the conversion of neutrons into protons (and vice versa) took place chiefly through collisions with electrons, positrons, neutrinos, and antineutrinos, not through the radioactive decay of neutrons. These points were noted in 1950 by C. Hayashi, and by 1953 Alpher and Herman (together with J. W. Follin, Jr.) had revised their model and carried out a substantially correct calculation of the shifting neutron-proton balance. This was, in fact, the first thoroughly modern analysis of the early history of the universe.

Nevertheless, no one in 1948 or 1953 set out to look for the predicted microwave radiation. Indeed, for years before 1965 it was not generally known to astrophysicists that in "big bang" models, the abundances of hydrogen and helium require the existence in the present universe of a cosmic radiation background, which might actually be observed. The surprising thing here is not so much that astrophysicists generally did not know of the prediction of Alpher and Herman—a paper or two can always sink out of sight in the great ocean of scientific literature. What is much more puzzling is that no one else pursued the same line of reasoning for over a decade. All the theoretical materials were at hand. It was not until 1964 that calculations of nucleosynthesis in a "big bang" model were begun again, by Ya. B. Zeldovich in Russia, Hoyle and R. J. Tayler in England, and Peebles in the U.S., all working independently. However, by this time Penzias and Wilson had already started their observations at Holmdel, and the discovery of the microwave background came about without any instigation by the cosmological theorists.

It is also puzzling that those who *did* know of the Alpher-Herman prediction did not seem to give it a great deal of emphasis. Alpher, Follin, and Herman themselves in their 1953 paper left the problem of nucleosynthesis for "future studies," so they were not in a position to recalculate the expected temperature of the microwave radiation background on the basis of their improved model. (Nor did they mention their earlier prediction that a 5° K radiation background was expected. They did report on some nucleosynthesis calculations at an American Physical Society meeting in 1953, but the three were moving to different laboratories and the work was never written up in final form.) Years later, in a letter to Penzias written after the discovery of the microwave radiation background, Gamow pointed out that in a 1953 article of his in the *Proceedings of the Royal Danish Academy,* he had predicted a radiation background with a temperature of 7° K, roughly the right order of magnitude. However, a glance at this 1953 paper shows that Gamow's prediction was based on a mathematically fallacious argument having to do with the age of the universe, and not on his own theory of cosmic nucleosynthesis.

It might be argued that the cosmic abundances of the light elements were not well enough known in the 1950s and early 1960s to draw any definite conclusions about the temperature of the radiation background. It is true that even now we are not really certain that there is a universal helium abundance in the range 20–30 percent. The important point, though, is that it has been believed since long before 1960 that most of the mass of the universe is in the form of hydrogen. (For instance, a 1956 survey by Hans Suess and Harold Urey gave a hydrogen abundance of 75 percent by weight.) And hydrogen is *not* produced in stars—it is the primitive fuel from which stars derive their energy by building up heavier elements. This is by itself enough to tell us that there must have been a large ratio of photons to nuclear particles, to prevent the cooking of all the hydrogen into helium and heavier elements in the early universe.

One may ask: When in fact did it become technologically possible to observe a 3° K isotropic radiation background? It is difficult to be precise about this, but my experimental colleagues tell me that the observation could have been made long before 1965, probably in the mid-1950s and perhaps even in the mid-1940s. In 1946 a team at the M.I.T. Radiation Laboratory, led by none other than Robert Dicke, was able to set an upper limit on any isotropic extraterrestrial radiation background: the equivalent temperature was less than 20° K at wavelengths 1.00, 1.25, and 1.50 centimeters. This measurement was a by-product of studies of atmospheric absorption, and was certainly not part of a program of observational cosmology. (In fact, Dicke informs me that by the time he started to wonder about a possible cosmic microwave radiation background, he had forgotten his own 20° K upper limit on the background temperature, obtained almost two decades earlier!)

It does not seem to me to be historically very important to pinpoint the moment when the detection of a 3° K isotropic microwave background became possible. The important point is that the radio astronomers did not know that they ought to try! In contrast, consider the history of the neutrino. When the neutrino was first hypothesized by Pauli in 1932, it was clear that there was not a shadow of a chance of observing it in any experiment then possible. However, the detection of neutrinos remained on physicists' minds as a challenging goal, and when nuclear reactors became available for such purposes in the 1950s the neutrino was sought and found. The contrast is even sharper with the case of the antiproton. After the positron was discovered in cosmic rays in 1932, theorists generally expected that the proton as well as the electron ought to have an antiparticle. There was no chance of producing antiprotons with the early cyclotrons available in the 1930s, but physicists remained aware of the problem, and in the 1950s an accelerator (the Bevatron at Berkeley) was built specifically to have enough energy to be able to produce antiprotons. Nothing of the sort happened in the case of the cosmic microwave radiation background, until Dicke and his associates set out to detect it in 1964. Even then, the Princeton group were not aware of the work of Gamow, Alpher, and Herman more than a decade earlier!

What, then, went wrong? It is possible here to trace at least three interesting reasons why the importance of a search for a 3° K microwave radiation background was not generally appreciated in the 1950s and early 1960s.

First, it must be understood that Gamow, Alpher, Herman, and Follin, et al., were working in the context of a broader cosmogonical theory. In their "big bang" theory essentially *all* complex nuclei, not only helium, were supposed to be built up in the early universe by a process of rapid addition of neutrons. However, although this theory correctly predicted the ratios of the abundances of some heavy elements, it ran into trouble in explaining why there are any heavy elements at all! As already mentioned, there is no stable nucleus with five or eight nuclear particles, so it is not possible to build nuclei heavier than helium by adding neutrons or protons to helium (He^4) nuclei, or by fusing pairs of helium nuclei. (This obstacle was first noted by Enrico Fermi and Anthony Turkevich.) Given this difficulty, it is easy to see why theorists were also unwilling even to take seriously the calculation of helium production in this theory.

The cosmological theory of element synthesis lost more ground as improvements were made in the alternative theory, that elements are synthesized in stars. In 1952 E. E. Salpeter showed that the gaps at nuclei with five or eight nuclear particles could be bridged in dense helium-rich stellar cores: collisions of two helium nuclei produce an unstable nucleus of beryllium (Be^8), and under these conditions of high density the beryllium nucleus may strike another helium nucleus before it decays, producing a stable carbon nucleus (C^{12}). (The density of the universe at the time of cosmological nucleosynthesis is much too low for this process to occur then.) In 1957 there appeared a famous paper by Geoffrey and Margaret Burbidge, Fowler, and Hoyle, which showed that the heavy elements could be built up in stars, particularly in stellar explosions such as supernovas, during periods of intense neutron flux. But even before the 1950s there was a powerful inclination among astrophysicists to believe that all elements other than hydrogen are produced in stars. Hoyle has remarked to me that this may have been an effect of the struggle that astronomers had to make in the early decades of this century to understand the source of the energy produced in stars. By 1940 the work of Hans Bethe and others had made it clear that the key process was the fusion of four hydrogen nuclei into one helium nucleus, and this picture had led in the 1940s and 1950s to rapid advances in the understanding of stellar evolution. As Hoyle says, after all these successes, it seemed to many astrophysicists to be perverse to doubt that the stars are the site of element formation.

But the stellar theory of nucleosynthesis also had its problems. It is difficult to see how stars could build up anything like a 25–30 percent helium abundance—indeed, the energy that would be released in this fusion would be much greater than stars seem to emit over their whole lifetime. The cosmological theory gets rid of this energy very nicely—it is simply lost in the general red shift. In 1964 Hoyle and R. J. Tayler pointed out that the large helium abundance of the present universe could not have been produced in ordinary stars, and they carried out a calculation of the amount of helium that would have been produced in the early stages of a "big bang," obtaining an abundance of 36 percent by weight. Oddly enough, they fixed the moment at which nucleosynthesis would have occurred at a more or less arbitrary temperature of 5,000 million degrees Kelvin, despite the fact that this assumption depends on the value chosen for a then unknown parameter, the ratio of photons to nuclear particles. Had they used their calculation to estimate this ratio from the *observed* helium abundance, they could have predicted a present microwave radiation background with a temperature roughly of the right order of magnitude. Nevertheless, it is striking that Hoyle, one of the originators of the steady-state theory, was willing to follow through this line of reasoning, and to acknowledge that it provided evidence for something like a "big bang" model.

Today it is generally believed that nucleosynthesis occurs *both* cosmologically and in stars; the helium and perhaps a few other light nuclei were synthesized in the early universe, while the stars are responsible for everything else. The "big bang" theory of nucleosynthesis, by trying to do too much, had lost the plausibility that it really deserved as a theory of helium synthesis.

Second, this was a classic example of a breakdown in communication between theorists and experimentalists. Most theorists never realized that an isotropic $3°$ K radiation background could ever be detected. In a letter to Peebles dated June 23, 1967, Gamow explained that neither he nor Alpher and Herman had considered the possibility of the detection of radiation left over from the "big bang," because at the time of their work on cosmology, radio astronomy was still in its childhood. (Alpher and Herman inform me, however, that they did in fact explore the possibility of observing the cosmic radiation background with radar experts at Johns Hopkins University, the Naval Research Laboratory, and the National Bureau of Standards, but were told that a radiation background temperature of $5°$ or $10°$ K was too low to be detected with the techniques then available.) On the other hand, some Soviet astrophysicists seem to have realized that a microwave background could be detected, but were led astray by the language in American technical journals. In a 1964 review article, Ya. B. Zeldovich carried out a correct calculation of the cosmic helium abundance for two possible values of the present radiation temperature, and correctly emphasized that these quantities are related because the number of photons per nuclear particle (or the entropy per nuclear particle) does not change with time. However, he seems to have been misled by the use of the term "sky temperature" in a 1961 article by E. A. Ohm in the *Bell System Technical Journal* to conclude that the radiation temperature had been measured to be less than 1° K. (The antenna used by Ohm was the same 20-foot horn reflector that was eventually used by Penzias and Wilson to discover the microwave background!) This, together with some rather low estimates of the cosmic helium abundance, led Zeldovich tentatively to abandon the idea of a hot early universe.

Of course, at the same time that information was flowing badly from experimenters to theorists, it was also flowing badly from theorists to experimenters. Penzias and Wilson had never heard of the Alpher-Herman prediction when they set out in 1964 to check their antenna.

Third, and I think most importantly, the "big bang" theory did not lead to a search for the 3° K microwave background because it was extraordinarily difficult for physicists to take seriously *any* theory of the early universe. (I speak here in part from recollections of my own attitude before 1965.) Every one of the difficulties mentioned above could have been overcome with a little effort. However, the first three minutes are so remote from us in time, the conditions of temperature and density are so unfamiliar, that we feel uncomfortable in applying our ordinary theories of statistical mechanics and nuclear physics.

This is often the way it is in physics—our mistake is not that we take our theories too seriously, but that we do not take them seriously enough. It is always hard to realize that these numbers and equations we play with at our desks have something to do with the real world. Even worse, there often seems to be a general agreement that certain phenomena are just not fit subjects for respectable theoretical and experimental effort. Gamow, Alpher, and Herman deserve tremendous credit above all for being willing to take the early universe seriously, for working out what known physical laws have to say about the first three minutes. Yet even they did not take the final step, to convince the radio astronomers that they ought to look for a microwave radiation background. The most important thing accomplished by the ultimate discovery of the 3° K radiation background in 1965 was to force us all to take seriously the idea that there *was* an early universe.

I have dwelt on this missed opportunity because this seems to me to be the most illuminating sort of history of science. It is understandable that so much of the historiography of science deals with its successes, with serendipitous discoveries, brilliant deductions, or the great magical leaps of a Newton or an Einstein. But I do not think it is possible really to understand the successes of science without understanding how *hard* it is—how easy it is to be led astray, how difficult it is to know at any time what is the next thing to be done.

VII

THE FIRST
ONE-HUNDREDTH SECOND

OUR ACCOUNT of the first three minutes in Chapter V did not begin at the beginning. Instead, we started at a "first frame" when the cosmic temperature had already cooled to 100,000 million degrees Kelvin, and the only particles present in large numbers were photons, electrons, neutrinos, and their corresponding antiparticles. If these really were the only types of particles in nature, we could perhaps extrapolate the expansion of the universe backward in time and infer that there must have been a real beginning, a state of infinite temperature and density, which occurred 0.0108 seconds before our first frame.

However, there are many other types of particles known to modern physics: muons, pi mesons, protons, neutrons, and so on. When we look back to earlier and earlier times, we encounter temperatures and densities so high that all of these particles would have been present in copious numbers in thermal equilibrium, and all in a state of continual mutual interaction. For reasons that I hope to make clear, we simply do not yet know enough about the physics of the elementary particles to be able to calculate the properties of such a mélange with any confidence. Thus, our ignorance of microscopic physics stands as a veil, obscuring our view of the very beginning.

Naturally, it is tempting to try to peek behind this veil. The temptation is particularly strong for theorists like myself, whose work has been much more in elementary particle physics than in astrophysics. Many of the interesting ideas in contemporary particle physics have such subtle consequences that they are extraordinarily difficult to test in laboratories today, but their consequences are quite dramatic when these ideas are applied to the very early universe.

The first problem we face in looking back to temperatures above 100,000 million degrees is presented by the "strong interactions" of elementary particles. The strong interactions are the forces that hold neutrons and protons together in an atomic nucleus. They are not familiar in everyday life, in the way that the electromagnetic and gravitational forces are, because their range is extremely short, about one ten million-millionth of a centimeter (10^{-13}cm). Even in molecules, whose nuclei are typically a few hundred millionths of a centimeter (10^{-8}cm) apart, the strong interactions between different nuclei have virtually no effect. However, as their name indicates, the strong interactions are very strong. When two protons are pushed close enough together, the strong interaction between them becomes about 100 times greater than the electrical repulsion; this is why the strong interactions are able to hold together atomic nuclei against the electrical repulsion of almost 100 protons. The explosion of a hydrogen bomb is caused by a rearrangement of neutrons and protons which allows them to be more tightly bound together by the strong interactions; the energy of the bomb is just the excess energy made available by this rearrangement.

It is the strength of the strong interactions that makes them so much more difficult to deal with mathematically than the electromagnetic interactions. When, for instance, we calculate the rate for scattering of two electrons due to the electromagnetic repulsion between them, we must add up an infinite number of contributions, each corresponding to a particular sequence of emission and absorption of photons and electron-positron pairs, symbolized in a "Feynman diagram" like those of figure 10. (The method of calculation that uses these diagrams was worked out in the late 1940s by Richard Feynman, then at Cornell. Strictly speaking, the rate for the scattering process is given by the *square* of a sum of contributions, one for each diagram.) Adding one more internal line to any diagram lowers the contribution of the diagram by a factor roughly equal to a fundamental constant of nature, known as the "fine structure constant." This constant is quite small, about 1/137.036. Complicated diagrams therefore give small contributions, and we can calculate the rate of the scattering process to an adequate approximation by adding up the contributions from just a few simple diagrams. (This is why we feel confident that we can predict atomic spectra with almost unlimited precision.) However, for the strong interactions, the constant that plays the role of the fine structure constant is roughly equal to one, not 1/137, and complicated diagrams therefore make just as large a contribution as simple dia-

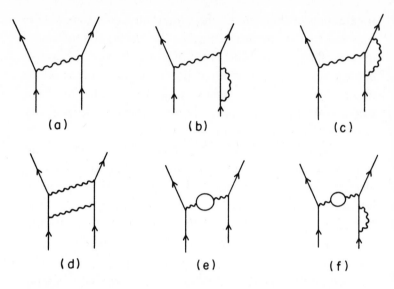

Figure 10. *Some Feynman Diagrams.* Shown here are some of the simpler Feynman diagrams for the process of electron-electron scattering. Straight lines denote electrons or positrons; wavy lines denote photons. Each diagram represents a certain numerical quantity which depends on the momenta and spins of the incoming and outgoing electrons; the rate of the scattering process is the square of the sum of these quantities, associated with all Feynman diagrams. The contribution of each diagram to this sum is proportional to a number of factors of 1/137 (the fine structure constant), given by the number of photon lines. Diagram (a) represents the exchange of a single photon and makes the leading contribution, proportional to 1/137. Diagrams (b), (c), (d), and (e) represent all the types of diagrams which make the dominant "radiative" corrections to (a); they all make contributions of order $(1/137)^2$. Diagram (f) makes an even smaller contribution, proportional to $(1/137)^3$.

grams. This problem, the difficulty of calculating rates for processes involving strong interactions, has been the single greatest obstacle to progress in elementary particle physics for the last quarter-century.

Not all processes involve strong interactions. The strong interactions affect only a class of particles known as "hadrons"; these include the nuclear particles and pi mesons, and other unstable particles known as K-mesons, eta mesons, lambda hyperons, sigma hyperons, and so on. The hadrons are generally heavier than the leptons (the name "lepton" is taken from the Greek word for "light"), but the really important difference between them is that hadrons feel the effects of the strong interactions, while the leptons—the neutrinos, electrons, and muons—do not. The fact that electrons do not feel the nuclear force is overwhelmingly important—together with the small mass of the electron, it is responsible for the fact that the cloud of electrons in an atom or a molecule is about 100,000 times larger than the atomic nuclei, and also that the chemical forces which hold atoms together in molecules are millions of times weaker than the forces which hold neutrons and protons together in nuclei. If the electrons in atoms and molecules felt the nuclear force, there would be no chemistry or crystallography or biology—only nuclear physics.

The temperature of 100,000 million degrees Kelvin, with which we began in Chapter V, was carefully chosen to be below the threshold temperature for all hadrons. (According to Table One on p. 156, the lightest hadron, the pi meson, has a threshold temperature of about 1.6 million million degrees Kelvin.) Thus, throughout the story told in Chapter V, the only particles present in large numbers were leptons and photons, and the interactions among them could safely be ignored.

How do we deal with higher temperatures, when hadrons and antihadrons would have been present in large numbers? There are two very different answers which reflect two very different schools of thought as to the nature of the hadrons.

According to one school, there really is no such thing as an "elementary" hadron. Every hadron is as fundamental as every other—not only stable and nearly stable hadrons like the proton and neutron, and not only moderately unstable particles like the pi mesons, K-mesons, eta meson, and hyperons, which live long enough to leave measurable tracks in photographic films or bubble chambers, but even totally unstable "particles" like the rho mesons, which live just long enough so that at a speed near that of light they can barely cross an atomic nucleus. This doctrine was developed in the late 1950s and early 1960s, particularly by Geoffrey Chew of Berkeley, and is sometimes known as "nuclear democracy."

With such a liberal definition of "hadron," there are literally hundreds of known hadrons whose threshold temperature is less than 100 million million degrees Kelvin, and probably hundreds more yet to be discovered. In some theories there is an unlimited number of species: the number of types of particles will increase faster and faster as we explore higher and higher masses. It might seem hopeless to try to make sense out of such a world, but the very complexity of the particle spectrum might lead to a kind of simplicity. For instance, the rho meson is a hadron that can be thought of as an unstable composite of two pi mesons; when we include rho mesons explicitly in our calculations, we are already to some extent taking account of the strong interaction between pi mesons; perhaps by including *all* hadrons explicitly in our thermodynamic calculations, we can ignore *all* other effects of the strong interactions.

Further, if there really is an unlimited number of species of hadron, then when we put more and more energy in a given volume the energy does not go into increasing the random speeds of the particles, but goes instead into increasing the numbers of types of particles present in the volume. The temperature then does not go up as fast with increasing energy density as it would if the number of hadron species were fixed. In fact, in such theories there can be a *maximum* temperature, a value of the temperature at which the energy density becomes infinite. This would be as insuperable an upper bound on the temperature as absolute zero is a lower bound. The idea of a maximum temperature in hadron physics is originally due to R. Hagedorn of the CERN laboratory in Geneva, and has been further developed by other theorists, including Kerson Huang of M. I. T. and myself. There is even a fairly precise estimate of what the maximum temperature would be—it is surprisingly low, about two million million degrees Kelvin (2×10^{12} ° K). As we look closer and closer to the beginning, the temperature would grow closer and closer to this maximum, and the variety of hadron types present would grow richer and richer. However, even under these exotic conditions there would still have been a beginning, a time of infinite energy density, roughly a hundredth of a second before the first frame of Chapter V.

There is another school of thought that is far more conventional, far closer to ordinary intuition than "nuclear democracy," and in my opinion also closer to the truth. According to this school, not all particles are equal; some really are elementary, and all the others are mere composites of the elementary particles. The elementary particles are thought to consist of the photon and all the known leptons, *but none of the known hadrons.* Rather, the hadrons are supposed to be composites of more fundamental particles, known as "quarks."

The original version of the quark theory is due to Murray Gell-Mann and (independently) George Zweig, both of Cal Tech. The poetic imagination of theoretical physicists has really run wild in naming the different sorts of quarks. The quarks come in different types, or "flavors," which are given names like "up," "down," "strange," and "charmed." Furthermore, each "flavor" of quark comes in three distinct "colors," which U.S. theorists usually call red, white, and blue. The small group of theoretical physicists in Peking has long favored a version of the quark theory, but they call them "stratons" instead of quarks because these particles represent a deeper stratum of reality than the ordinary hadrons.

If the quark idea is right, then the physics of the very early universe may be simpler than was thought. It is possible to infer something about the forces between quarks from their spatial distribution inside a nuclear particle, and this distribution can in turn be determined (if the quark model is true) from observations of high-energy collisions of electrons with nuclear particles. In this way, it was found a few years ago by an M.I.T.-Stanford Linear Accelerator Center collaboration that the force between quarks seems to disappear when the quarks are very close to each other. This suggests that at some temperature, around several million million degrees Kelvin, the hadrons would simply break up into their constituent quarks, just as atoms break up into electrons and nuclei at a few thousand degrees, and nuclei break up into protons and neutrons at a few thousand million degrees. According to this picture, at very early times the universe could be considered to consist of photons, leptons, antileptons, quarks, and antiquarks, all moving essentially as free particles, and each particle species therefore in effect furnishing just one more kind of black-body radiation. It is easy then to calculate that there must have been a beginning, a state of infinite density *and* infinite temperature, about one-hundredth of a second before the first frame.

These rather intuitive ideas have recently been put on a much firmer mathematical foundation. In 1973 it was shown by three young theorists, Hugh David Politzer of Harvard and David Gross and Frank Wilczek of Princeton, that, in a special class of quantum field theories, the forces between quarks do actually become weaker as the quarks are pushed closer together. (This class of theories is known as the "non-Abelian gauge theories" for reasons too technical to explain here.) These theories have the remarkable property of "asymptotic freedom": at asymptotically short distances or high energies, the quarks behave as free particles. It has even been shown by J. C. Collins and M. J. Perry at the University of Cambridge, that in any asymptotically free theory, the properties of a medium at sufficiently high temperature and density are essentially the same as if the medium consisted purely of free particles. The asymptotic freedom of these non-Abelian gauge theories thus provides a solid mathematical justification for the very simple picture of the first hundredth of a second— that the universe was made up of free elementary particles.

The quark model works very well in a wide variety of applications. Protons and neutrons really do behave as if they consist of three quarks, rho mesons behave as if they consist of a quark and an antiquark, and so on. But despite this success, the quark model presents us with a great puzzle: even with the highest energies available from existing accelerators, it has so far proved impossible to break up any hadron into its constituent quarks.

The same inability to isolate free quarks also appears in cosmology. If hadrons really broke up into free quarks under the conditions of high temperature that prevailed in the early universe, then one might expect some free quarks to be left over at the present time. The Soviet astrophysicist Ya. B. Zeldovich has estimated that free leftover quarks should be roughly as common in the present universe as gold atoms. Needless to say, gold is not abundant, but an ounce of gold is a good deal easier to purchase than an ounce of quarks.

F

The puzzle of the nonexistence of isolated free quarks is one of the most important problems facing theoretical physics at the present moment. It has been suggested by Gross and Wilczek and by myself that "asymptotic freedom" provides a possible explanation. If the strength of the interaction between two quarks decreases as they are pushed close together, it also increases as they are pulled farther apart. The energy required to pull a quark away from the other quarks in an ordinary hadron therefore increases with increasing distance, and it seems eventually to become great enough to create new quark-antiquark pairs out of the vacuum. In the end, one winds up not with several free quarks, but with several ordinary hadrons. It is exactly like trying to isolate one end of a piece of string: if you pull very hard, the string will break, but the final result is two pieces of string, each with two ends! Quarks were close enough together in the early universe so that they did not feel these forces, and could behave like free particles. However, *every* free quark present in the very early universe must, as the universe expanded and cooled, have either annihilated with an antiquark or else found a resting place inside a proton or neutron.

So much for the strong interactions. There are further problems in store for us as we turn the clock back toward the very beginning.

One truly fascinating consequence of modern theories of elementary particles is that the universe may have suffered a *phase transition*, like the freezing of water when it falls below $273°$ K ($= 0°C$). This phase transition is associated not with the strong interactions, but with the other class of short-range interaction in particle physics, the *weak* interactions.

The weak interactions are those responsible for certain radioactive decay processes like the decay of a free neutron (see

p. 93) or, more generally, for any reaction involving a neutrino (see p. 98). As their name indicates, the weak interactions are much weaker than the electromagnetic or strong interactions. For instance, in a collision between a neutrino and an electron at an energy of one million electron volts, the weak force is about one ten-millionth (10^{-7}) of the electromagnetic force between two electrons colliding at the same energy.

Despite the weakness of the weak interactions, it has long been thought that there might be a deep relation between the weak and electromagnetic forces. A field theory which unifies these two forces was proposed in 1967 by myself, and independently in 1968 by Abdus Salam. This theory predicted a new class of weak interactions, the so-called neutral currents, whose existence was confirmed experimentally in 1973. It received further support from the discovery, starting in 1974, of a whole family of new hadrons. The key idea of this kind of theory is that nature has a very high degree of symmetry, which relates the various particles and forces, but which is obscured in ordinary physical phenomena. The field theories used since 1973 to describe the strong interactions are of the same mathematical type (non-Abelian gauge theories), and many physicists now believe that gauge theories may provide a unified basis for understanding all the forces of nature: weak, electromagnetic, strong, and perhaps gravitational forces.

For studies of the early universe, the important point about the gauge theories is that, as pointed out in 1972 by D. A. Kirzhnits and A. D. Linde of the Lebedev Physical Institute in Moscow, these theories exhibit a phase transition, a kind of freezing, at a "critical temperature" of about 3,000 million million degrees (3×10^{15} ° K). At temperatures below the critical temperature, the universe was as it is now: weak interactions were weak, and of short range. At temperatures above the critical temperature, the essential unity between the weak and electromagnetic interactions was manifest: the weak interactions obeyed the same sort of inverse-square law as the electromagnetic interactions, and had about the same strength.

The analogy with a freezing glass of water is instructive here. Above the freezing point, liquid water exhibits a high degree of homogeneity: the probability of finding a water molecule at one point inside the glass is just the same as at any other point. However, when the water freezes this symmetry among different points in space is partly lost: the ice forms a crystal lattice, with water molecules occupying certain regularly spaced positions, and with almost zero probability of finding water molecules anywhere else. In the same way, when the universe "froze" as the temperature fell below 3,000 million million degrees, a symmetry was lost—not its spatial homogeneity, as in our glass of ice, but the symmetry between the weak and the electromagnetic interactions.

It may be possible to carry the analogy even farther. As everyone knows, when water freezes it does not usually form a perfect crystal of ice, but something much more complicated: a great mess of crystal domains, separated by various types of crystal irregularities. Did the universe also freeze into domains? Do we live in one such domain, in which the symmetry between the weak and electromagnetic interactions has been broken in a particular way, and will we eventually discover other domains?

So far our imagination carried us back to a temperature of 3,000 million million degrees, and we have had to deal with the strong, weak, and electromagnetic interactions. What about the one other grand class of interactions known to physics, the gravitational interactions? Gravitation has of course played an important role in our story, because it controls the relation between the density of the universe and its rate of expansion. However, gravity has not yet been found to have any effect on the *internal* properties of any part of the early universe. This is because of the extreme weakness of the gravitational force; for instance, the gravitational force between the electron and the proton in a hydrogen atom is weaker than the electrical force by 39 powers of 10.

(One illustration of the weakness of gravitation in cosmological processes is provided by the process of particle production in gravitational fields. It has been pointed out by Leonard Parker of the University of Wisconsin that the "tidal" effects of the gravitational field of the universe would have been great enough, at a time about one million million million-millionth of a second (10^{-24} sec) after the beginning, to produce particle-antiparticle pairs out of empty space. However, gravitation was still so weak at these temperatures that the number of particles produced in this way made a negligible contribution to the particles already present in thermal equilibrium.)

Nevertheless, we can at least imagine a time when gravitational forces would have been as strong as the strong nuclear interactions discussed above. Gravitational fields are generated not only by particle masses, but by all forms of energy. The earth is going around the sun a little faster than it otherwise would if the sun were not hot, because the energy in the sun's heat adds a little to the source of its gravitation. At super-high temperatures the energies of particles in thermal equilibrium can become so large that the gravitational forces between them become as strong as any other forces. We can estimate that this state of affairs was reached at a temperature of about 100 million million million million million degrees (10^{32} ° K).

At this temperature all sorts of strange things would have

been going on. Not only would gravitational forces have been strong and particle production by gravitational fields copious—the very idea of "particle" would not yet have had any meaning. The "horizon," the distance beyond which it is impossible yet to have received any signals (see p. 41), would at this time be closer than one wavelength of a typical particle in thermal equilibrium. Speaking loosely, each particle would be about as big as the observable universe!

We do not know enough about the quantum nature of gravitation even to speculate intelligently about the history of the universe before this time. We can make a crude estimate that the temperature of 10^{32} ° K was reached some 10^{-43} seconds after the beginning, but it is not really clear that this estimate has any meaning. Thus, whatever other veils may have been lifted, there is one veil, at a temperature of 10^{32} ° K, that still obscures our view of the earliest times.

However, none of these uncertainties make much difference to the astronomy of A.D. 1976. The point is that during the whole of the first second the universe was presumably in a state of thermal equilibrium, in which the numbers and distributions of all particles, even neutrinos, were determined by the laws of statistical mechanics, not by the details of their prior history. When we measure the abundance today of helium, or microwave radiation, or even of neutrinos, we are observing the relics of the state of thermal equilibrium which ended at the close of the first second. As far as we know, nothing that we can observe depends on the history of the universe prior to that time. (In particular, nothing we now observe depends on whether the universe was isotropic and homogeneous before the first second, except perhaps the photon-to-nuclear-particle ratio itself.) It is as if a dinner were prepared with great care—the freshest ingredients, the most carefully chosen spices, the finest wines—and then thrown all together in a great pot to boil for a few hours. It would be difficult for even the most discriminating diner to know what he was being served.

There is one possible exception. The phenomenon of gravitation, like that of electromagnetism, can be manifested in the form of waves as well as in the more familiar form of a static action at a distance. Two electrons at rest will repel each other with a static electric force that depends on the distance between them, but if we wiggle one electron back and forth, the other will not feel any change in the force acting on it until there is time for news of the change in separation to be carried on an electromagnetic wave from one particle to the other. It hardly needs to be said that these waves travel at the speed of light—they *are* light, although not necessarily visible light. In the same way, if some ill-advised giant were to wiggle the sun back and forth, we on earth would not feel the effect for eight minutes, the time required for a wave to travel at the speed of light from the sun to the earth. This is *not* a light wave, a wave of oscillating electric and magnetic fields, but rather a gravitational wave, in which the oscillation is in the gravitational fields. Just as for electromagnetic waves, we lump together gravitational waves of all wavelengths under the term "gravitational radiation."

Gravitational radiation interacts far more weakly with matter than electromagnetic radiation, or even neutrinos. (For this reason, although we are reasonably confident on theoretical grounds of the existence of gravitational radiation, the most strenuous efforts have so far apparently failed to detect gravitational waves from any source.) Gravitational radiation would therefore have gone out of thermal equilibrium with the other contents of the universe very early—in fact, when the temperature was about 10^{32} ° K. Since then, the effective temperature of the gravitational radiation has simply dropped in inverse proportion to the size of the universe. This is just the same law of decrease as obeyed by the temperature of the rest of the contents of the universe, except that the annihilation of quark-antiquark and lepton-antilepton pairs has heated the rest of the universe but not the gravitational radiation. Therefore, the universe today should be filled with gravitational radiation at a temperature similar to but somewhat less than that of the neutrinos or photons—perhaps about 1° K. Detection of this radiation would represent a direct observation of the very earliest moment in the history of the universe that can even be contemplated by present-day theoretical physics. Unfortunately there does not seem to be the slightest chance of detecting a 1° K background of gravitational radiation in the foreseeable future.

With the aid of a good deal of highly speculative theory, we have been able to extrapolate the history of the universe back in time to a moment of infinite density. But this leaves us unsatisfied. We naturally want to know what there was before this moment, before the universe began to expand and cool.

One possibility is that there never really was a state of infinite density. The present expansion of the universe may have begun at the end of a previous age of contraction, when the density of the universe had reached some very high but finite value. I will have a little more to say about this possibility in the next chapter.

However, although we do not know that it is true, it is at least logically possible that there *was* a beginning, and that time itself has no meaning before that moment. We are all used to the idea of an absolute zero of temperature. It is impossible to cool anything below −273.16° C, not because it is too hard or because no one has thought of a sufficiently clever refrigerator, but because temperatures lower than absolute zero just have no meaning—we cannot have less heat than no heat at all. In the same way, we may have to get used to the idea of an absolute zero of time—a moment in the past beyond which it is in principle impossible to trace any chain of cause and effect. The question is open, and may always remain open.

To me, the most satisfying thing that has come out of these speculations about the very early universe is the possible parallel between the history of the universe and its logical structure. Nature now exhibits a great diversity of types of particles and types of interactions. Yet we have learned to look beneath this diversity, to try to see the various particles and interactions as aspects of a simple unified gauge field theory. The present universe is so cold that the symmetries among the different particles and interactions have been obscured by a kind of freezing; they are not manifest in ordinary phenomena, but have to be expressed mathematically, in our gauge field theories. That which we do now by mathematics was done in the very early universe by heat—physical phenomena directly exhibited the essential simplicity of nature. But no one was there to see it.

VIII

EPILOGUE:
THE PROSPECT AHEAD

THE UNIVERSE will certainly go on expanding for a while. As to its fate after that, the standard model gives an equivocal prophecy: It all depends on whether the cosmic density is less or greater than a certain critical value.

As we saw in Chapter II, if the cosmic density is *less* than the critical density, then the universe is of infinite extent and will go on expanding forever. Our descendants, if we have any then, will see thermonuclear reactions slowly come to an end in all the stars, leaving behind various sorts of cinder: black dwarf stars, neutron stars, perhaps black holes. Planets may continue in orbit, slowing down a little as they radiate gravitational waves but never coming to rest in any finite time. The cosmic backgrounds of radiation and neutrinos will continue to fall in temperature in inverse proportion to the size of the universe, but they will not be missed; even now we can barely detect the 3° K microwave radiation background.

On the other hand, if the cosmic density is *greater* than the critical value, then the universe is finite and its expansion will eventually cease, giving way to an accelerating contraction. If, for instance, the cosmic density is twice its critical value, and if the presently popular value of the Hubble constant (15 kilometers per second per million light years) is correct, then the universe is now 10,000 million years old; it will go on expanding for another 50,000 million years, and then begin to contract. (See figure 4, p. 38.) The contraction is just the expansion run backward: after 50,000 million years the universe would have regained its present size, and after another 10,000 million years it would approach a singular state of infinite density.

During at least the early part of the contracting phase, astronomers (if there are any) will be able to amuse themselves by observing both red shifts and blue shifts. Light from nearby galaxies would have been emitted at a time when the universe was larger than it is when the light is observed, so when it is observed this light will appear to be shifted toward the short wavelength end of the spectrum, i.e., toward the blue. On the other hand, the light from extremely distant objects would have been emitted at a time when the universe was still in the early stages of its expansion, when the universe was even smaller than it is when the light is observed, so when it is observed this light will appear to be shifted toward the long wavelength end of the spectrum, i.e., toward the red.

The temperature of the cosmic backgrounds of photons and neutrinos will fall and then rise as the universe expands and then contracts, always in inverse proportion to the size of the universe. If the cosmic density now is twice its critical value, then our calculations show that the universe at its maximum dilation will be just twice as large as at present, so the microwave background temperature will then be just one-half its present value of 3° K, or about 1.5° K. Then, as the universe begins to contract, the temperature will start to rise.

At first no alarms will sound—for thousands of millions of years the radiation background will be so cool that it would take a great effort to detect it at all. However, when the universe has recontracted to one-hundredth its present size, the radiation background will begin to dominate the sky: the night sky will be as warm (300° K) as our present sky at day. Seventy million years later the universe will have contracted another tenfold, and our heirs and assigns (if any) will find the sky intolerably bright. Molecules in planetary and stellar atmospheres and in interstellar space will begin to dissociate into their constituent atoms, and the atoms will break up into free electrons and atomic nuclei. After another 700,000 years, the cosmic temperature will be at ten million degrees; then stars and planets themselves will dissolve into a cosmic soup of radiation, electrons, and nuclei. The temperature will rise to ten thousand million degrees in another 22 days. The nuclei will then begin to break up into their constituent protons and neutrons, undoing all the work of both stellar and cosmological nucleosynthesis. Soon after that, electrons and positrons will be created in great numbers in photon-photon collisions, and the cosmic background of neutrinos and antineutrinos will regain thermal communion with the rest of the universe.

Can we really carry this sad story all the way to its end, to a state of infinite temperature and density? Does time really have a stop some three minutes after the temperature reaches a thousand million degrees? Obviously, we cannot be sure. All the uncertainties that we met in the preceding chapter, in trying to explore the first hundredth of a second, will return to perplex us as we look into the last hundredth of a second. Above all, the whole universe must be described in the language of quantum mechanics at temperatures above 100 million million million million million degrees (10^{32} ° K), and no one has any idea what happens then. Also, if the universe is not really isotropic and homogeneous (see the end of Chapter V), then our whole story may have lost its validity long before we would have to face the problems of quantum cosmology.

From these uncertainties some cosmologists derive a sort of hope. It may be that the universe will experience a kind of cosmic "bounce," and begin to reexpand. In the *Edda*, after the final battle of the gods and giants at Ragnorak, the earth is destroyed by fire and water, but the waters recede, the sons of Thor come up from Hell carrying their father's hammer, and the whole world begins once more. But if the universe does reexpand, its expansion will again slow to a halt and be followed by another contraction, ending in another cosmic Ragnorak, followed by another bounce, and so on forever.

If this is our future, it presumably also is our past. The present expanding universe would be only the phase following the last contraction and bounce. (Indeed, in their 1965 paper on the cosmic microwave radiation background, Dicke, Peebles, Roll, and Wilkinson assumed that there was a previous complete phase of cosmic expansion and contraction, and they argued that the universe must have contracted enough to raise the temperature to at least ten thousand million degrees in order to break up the heavy elements formed in the previous phase.) Looking farther back, we can imagine an endless cycle of expansion and contraction stretching into the infinite past, with no beginning whatever.

Some cosmologists are philosophically attracted to the oscillating model, especially because, like the steady-state model, it nicely avoids the problem of Genesis. It does, however, face one severe theoretical difficulty. In each cycle the ratio of photons to nuclear particles (or, more precisely, the entropy per nuclear particle) is slightly increased by a kind of friction (known as "bulk viscosity") as the universe expands and contracts. As far as we know, the universe would then start each new cycle with a new, slightly larger ratio of photons to nuclear particles. Right now this ratio is large, but not infinite, so it is hard to see how the universe could have previously experienced an infinite number of cycles.

However all these problems may be resolved, and whichever cosmological model proves correct, there is not much of comfort in any of this. It is almost irresistible for humans to believe that we have some special relation to the universe, that human life is not just a more-or-less farcical outcome of a chain of accidents reaching back to the first three minutes, but that we were somehow built in from the beginning. As I write this I happen to be in an airplane at 30,000 feet, flying over Wyoming en route home from San Francisco to Boston. Below, the earth looks very soft and comfortable—fluffy clouds here and there, snow turning pink as the sun sets, roads stretching straight across the country from one town to another. It is very hard to realize that this all is just a tiny part of an overwhelmingly hostile universe. It is even harder to realize that this present universe has evolved from an unspeakably unfamiliar early condition, and faces a future extinction of endless cold or intolerable heat. The more the universe seems comprehensible, the more it also seems pointless.

But if there is no solace in the fruits of our research, there is at least some consolation in the research itself. Men and women are not content to comfort themselves with tales of gods and giants, or to confine their thoughts to the daily affairs of life; they also build telescopes and satellites and accelerators, and sit at their desks for endless hours working out the meaning of the data they gather. The effort to understand the universe is one of the very few things that lifts human life a little above the level of farce, and gives it some of the grace of tragedy.

TABLES

One. *Properties of Some Elementary Particles*

	Particle	Symbol	Rest energy (million electron volts)	Threshold temperature (thousand million degrees K)	Effective number of species	Mean life (seconds)
	Photon	γ	0	0	$1 \times 2 \times 1 = 2$	stable
Leptons	Neutrinos	$\nu_e, \bar{\nu}_e$	0	0	$2 \times 1 \times 7/8 = 7/4$	stable
		$\nu_\mu, \bar{\nu}_\mu$	0	0	$2 \times 1 \times 7/8 = 7/4$	stable
	Electron	e^-, e^+	0.5110	5.930	$2 \times 2 \times 7/8 = 7/2$	stable
	Muon	μ^-, μ^+	105.66	1226.2	$2 \times 2 \times 7/8 = 7/2$	2.197×10^{-6}
Hadrons	Pi mesons	π^0	134.96	1566.2	$1 \times 1 \times 1 = 1$	0.8×10^{-16}
		π^+, π^-	139.57	1619.7	$2 \times 1 \times 1 = 2$	2.60×10^{-8}
	Proton	$p, \bar{p}$	938.26	10,888	$2 \times 2 \times 7/8 = 7/2$	stable
	Neutron	$n, \bar{n}$	939.55	10,903	$2 \times 2 \times 7/8 = 7/2$	920

Properties of Some Elementary Particles. The "rest energy" is the energy that would be released if all the mass of the particle were converted into energy. The "threshold temperature" is the rest energy divided by Boltzmann's constant; it is the temperature above which a particle can be freely created out of thermal radiation. The "effective number of species" gives the relative contribution of each type of particle to the total energy, pressure, and entropy, at temperatures high above the threshold temperature. This number is written as the product of three factors: the first factor is 2 or 1 according to whether the particle does or does not have a distinct antiparticle; the second factor is the number of possible orientations of the particle's spin; the last factor is ⅞ or 1 according to whether or not the particle obeys the Pauli Exclusion Principle. The "mean life" is the average length of time the particle survives before it suffers a radioactive decay into other particles.

Two. *Properties of Some Kinds of Radiation*

	Wavelength (centimeters)	Photon energy (electron volts)	Black-body temperature (degrees Kelvin)
Radio (up to VHF)	> 10	< 0.00001	< 0.03
Microwave	0.01 to 10	0.00001 to 0.01	0.03 to 30
Infrared	0.0001 to 0.01	0.01 to 1	30 to 3,000
Visible	2×10^{-5} to 10^{-4}	1 to 6	3,000 to 15,000
Ultraviolet	10^{-7} to 2×10^{-5}	6 to 1,000	15,000 to 3,000,000
X ray	10^{-9} to 10^{-7}	1,000 to 100,000	3×10^6 to 3×10^8
γ ray	$< 10^{-9}$	$> 100,000$	$> 3 \times 10^8$

Properties of Some Kinds of Radiation. Each kind of radiation is characterized by a certain range of wavelengths, given here in centimeters. Corresponding to this range of wavelengths is a range of photon energies, given here in electron volts. The "black-body temperature" is the temperature at which black-body radiation would have most of its energy concentrated near the given wavelengths; this temperature is given here in degrees Kelvin. (For instance, the wavelength to which Penzias and Wilson were tuned in their discovery of the cosmic radiation background was 7.35 cm., so this is microwave radiation; the photon energy released when a nucleus undergoes a radioactive transmutation is typically about a million electron volts, so this is a γ ray; and the surface of the sun is at a temperature of 5800° K, so the sun emits visible light.) Of course, the divisions between the different kinds of radiation are not perfectly sharp, and there is no universal agreement on the various wavelength ranges.

Glossary

ABSOLUTE LUMINOSITY   The total energy emitted per unit time by any astronomical body.

ANDROMEDA NEBULA   The large galaxy nearest to our own. A spiral, containing about 3×10^{11} solar masses. Listed as M31 in the Messier catalog, NGC 224 in the "New General Catalog."

ANGSTROM UNIT   One hundred-millionth of a centimeter (10^{-8} cm). Denoted by Å. Typical atomic sizes are a few Angstroms; typical wavelengths of visible light are a few thousand Angstroms.

ANTIPARTICLE   A particle with the same mass and spin as another particle, but with equal and opposite electric charge, baryon number, lepton number, and so on. To every particle there is a corresponding antiparticle, except that certain purely neutral particles like the photon and π^0 meson are their own antiparticles. The *antineutrino* is the antiparticle of the neutrino; the *antiproton* is the antiparticle of the proton; and so on. *Antimatter* consists of the antiprotons, antineutrons, and antielectrons, or positrons.

APPARENT LUMINOSITY   The total energy received per unit time and per unit receiving area from any astronomical body.

ASYMPTOTIC FREEDOM   The property of some field theories of the strong interactions, that the forces become increasingly weak at short distances.

BARYONS   A class of strongly interacting particles, including neutrons, protons, and the unstable hadrons known as hyperons. *Baryon number* is the total number of baryons present in a system, minus the total number of antibaryons.

"BIG BANG" COSMOLOGY   The theory that the expansion of the universe began at a finite time in the past, in a state of enormous density and pressure.

BLACK-BODY RADIATION   Radiation with the same energy density in each wavelength range as the radiation emitted from a totally absorbing heated body. The radiation in any state of thermal equilibrium is black-body radiation.

BLUE SHIFT   The shift of spectral lines toward shorter wavelengths, caused by the Doppler effect for an approaching source.

G

Glossary

BOLTZMANN'S CONSTANT   The fundamental constant of statistical mechanics, which relates the temperature scale to units of energy. Usually denoted k, or k_B. Equal to 1.3806×10^{-16} ergs per degree Kelvin, or 0.00008617 electron volts per degree Kelvin.

CEPHEID VARIABLES   Bright variable stars, with a well-defined relation among absolute luminosity, period of variability, and color. Named after the star δ Cephei in the constellation Cepheus ("the King"). Used as indicators of distance for relatively near galaxies.

CHARACTERISTIC EXPANSION TIME   Reciprocal of the Hubble constant. Roughly, 100 times the time in which the universe would expand by 1 percent.

CONSERVATION LAW   A law which states that the total value of some quantity does not change in any reaction.

COSMIC RAYS   High-energy charged particles which enter our earth's atmosphere from outer space.

COSMOLOGICAL CONSTANT   A term added by Einstein in 1917 to his gravitational field equations. Such a term would produce a repulsion at very large distances, and would be needed in a static universe to balance the attraction due to gravitation. There is no reason at present to suspect the existence of a cosmological constant.

COSMOLOGICAL PRINCIPLE   The hypothesis that the universe is isotropic and homogeneous.

CRITICAL DENSITY   The minimum present cosmic mass density required if the expansion of the universe is eventually to cease and be succeeded by a contraction. The universe is spatially finite if the cosmic density exceeds the critical density.

CRITICAL TEMPERATURE   The temperature at which a phase transition occurs.

CYANOGEN   The chemical compound CN, formed of carbon and nitrogen. Found in interstellar space by absorption of visible light.

DECELERATION PARAMETER   A number which characterizes the rate at which the recession of distant galaxies is slowing down.

DENSITY   The amount of any quantity per unit volume. The *mass density* is the mass per unit volume; this is often simply referred to as "the density." The *energy density* is the energy per unit volume; the *number density* or *particle density* is the number of particles per unit volume.

DEUTERIUM   A heavy isotope of hydrogen, H^2. The nuclei of deuterium, called *deuterons*, consist of one proton and one neutron.

DOPPLER EFFECT   The change in frequency of any signal, caused by a relative motion of source and receiver.

ELECTRON   The lightest massive elementary particle. All chemical properties of atoms and molecules are determined by the electrical interactions of electrons with each other and with the atomic nuclei.

ELECTRON VOLT A unit of energy, convenient in atomic physics, equal to the energy acquired by one electron in passing through a voltage difference of one volt. Equal to 1.60219×10^{-12} ergs.

ENTROPY A fundamental quantity of statistical mechanics, related to the degree of disorder of a physical system. The entropy is conserved in any process in which thermal equilibrium is continually maintained. The second law of thermodynamics says that the total entropy never decreases in *any* reaction.

ERG The unit of energy in the centimeter-gram-second system. The kinetic energy of a mass of one gram traveling at one centimeter per second is one-half erg.

FEYNMAN DIAGRAMS Diagrams which symbolize various contributions to the rate of an elementary particle reaction.

FINE STRUCTURE CONSTANT Fundamental numerical constant of atomic physics and quantum electrodynamics, defined as the square of the charge of the electron divided by the product of Planck's constant and the speed of light. Denoted α. Equal to $1/137.036$.

FREQUENCY The rate at which crests of any sort of wave pass a given point. Equal to the speed of the wave divided by the wavelength. Measured in cycles per second, or "Hertz."

FRIEDMANN MODEL The mathematical model of the space-time structure of the universe, based on general relativity (without a cosmological constant) and the Cosmological Principle.

GALAXY A large gravitationally bound cluster of stars, containing up to 10^{12} solar masses. Our galaxy is sometimes called "The Galaxy." Galaxies are generally classified according to their shape, as elliptical, spiral, barred spiral, or irregular.

GAUGE THEORIES A class of field theories currently under intense study as possible theories of the weak, electromagnetic, and strong interactions. Such theories are invariant under a symmetry transformation, whose effect varies from point to point in space-time. The term "gauge" comes from the ordinary English word meaning "measure," but the term is used mostly for historical reasons.

GENERAL RELATIVITY The theory of gravitation developed by Albert Einstein in the decade 1906–1916. As formulated by Einstein, the essential idea of general relativity is that gravitation is an effect of the curvature of the space-time continuum.

GRAVITATIONAL WAVES Waves in the gravitational field, analogous to the light waves in the electromagnetic field. Gravitational waves travel at

Glossary

the same speed as light waves, 299,792 kilometers per second. There is no generally accepted experimental evidence for gravitational waves, but their existence is required by general relativity, and is not seriously in doubt. The quantum of gravitational radiation, analogous to the photon, is called the *graviton*.

HADRON Any particle that participates in the strong interaction. Hadrons are divided into baryons (such as the neutron and proton), which obey the Pauli Exclusion Principle, and mesons, which do not.

HELIUM The second lightest, and second most abundant, chemical element. There are two stable isotopes of helium: the nucleus of He^4 contains two protons and two neutrons, while the nucleus of He^3 contains two protons and one neutron. Atoms of helium contain two electrons outside the nucleus.

HOMOGENEITY The assumed property of the universe, that at a given time it appears the same to all typical observers, wherever located.

HORIZON In cosmology, the distance from beyond which no light signal would have yet had time to reach us. If the universe has a definite age, then the distance to the horizon is of the order of the age times the speed of light.

HUBBLE'S LAW The relation of proportionality between the velocity of recession of moderately distant galaxies and their distance. The *Hubble constant* is the ratio of velocity to distance in this relation, and is denoted H or H_0.

HYDROGEN The lightest and most abundant chemical element. The nucleus of ordinary hydrogen consists of a single proton. There are also two heavier isotopes, deuterium and tritium. Atoms of any sort of hydrogen consist of a hydrogen nucleus and a single electron; in positive *hydrogen ions* the electron is missing.

HYDROXYL ION The ion OH^-, formed of an oxygen atom, a hydrogen atom, and one extra electron.

INFRARED RADIATION Electromagnetic waves with wavelength between about 0.0001 cm and 0.01 cm (ten thousand to one million Angstroms), intermediate between visible light and microwave radiation. Bodies at room temperature radiate chiefly in the infrared.

ISOTROPY The assumed property of the universe, that to a typical observer it looks the same in all directions.

JEANS MASS The minimum mass for which gravitational attraction can overcome internal pressure and produce a gravitationally bound system. Denoted M_J.

KELVIN The temperature scale, like the Centigrade scale, but with absolute zero instead of the melting point of ice as the zero of temperature. The melting point of ice at a pressure of one atmosphere is 273.15° K.

LEPTON   A class of particles which do not participate in the strong interactions, including the electron, muon, and neutrino. *Lepton Number* is the total number of leptons present in a system, minus the total number of antileptons.

LIGHT YEAR   The distance that a light ray travels in one year, equal to 9.4605 million million kilometers.

MAXIMUM TEMPERATURE   The upper limit to temperature, implied by certain theories of the strong interactions. Estimated in these theories as two million million degrees Kelvin.

MEAN FREE PATH   The average distance traveled by a given particle between collisions with the medium in which it moves. The *mean free time* is the average time between collisions.

MESONS   A class of strongly interacting particles, including the pi mesons, K-mesons, rho mesons, and so on, with zero baryon number.

MESSIER NUMBERS   The catalog numbers of various nebulae and star clusters in the listing of Charles Messier. Usually abbreviated as M . . . ; thus the Andromeda Nebula is M31.

MICROWAVE RADIATION   Electromagnetic waves with wavelength between about 0.01 cm and 10 cm, intermediate between very-high-frequency radio and infrared radiation. Bodies with temperatures of a few degrees Kelvin radiate chiefly in the microwave band.

MILKY WAY   The ancient name of the band of stars which mark the plane of our galaxy. Sometimes used as a name for our galaxy itself.

MUON   An unstable elementary particle of negative charge, similar to the electron but 207 times heavier. Denoted μ. Sometimes called *mu meson*, but not strongly interacting like true mesons.

NEBULAE   Extended astronomical objects with a cloudlike appearance. Some nebulae are galaxies; others are actual clouds of dust and gas within our galaxy.

NEUTRINO   A massless electrically neutral particle, having only weak and gravitational interactions. Denoted ν. Neutrinos come in at least two varieties, known as electron-type (ν_e) and muon-type (ν_μ).

NEWTON'S CONSTANT   The fundamental constant of Newton's and Einstein's theories of gravitation. Denoted G. In Newton's theory the gravitational force between two bodies is G times the product of the masses divided by the square of the distance between them. In metric units, equal to 6.67×10^{-8} cm³/gm sec.

NUCLEAR DEMOCRACY   The doctrine that all hadrons are equally fundamental.

NUCLEAR PARTICLES   The particles, protons and neutrons, found in the nuclei of ordinary atoms. Usually shortened to *nucleons*.

PARSEC   Astronomical unit of distance. Defined as distance of an object

Glossary

whose *parallax* (annual shift in sky due to earth's motion around sun) is one second of arc. Abbreviated pc. Equal to 3.0856×10^{13} kilometers, or 3.2615 light years. Generally used in astronomical literature in preference to light years. Conventional unit of cosmology is one million parsecs, or *megaparsec*, abbreviated Mpc. Hubble's constant is usually given in kilometers per second per megaparsec.

PAULI EXCLUSION PRINCIPLE   The principle that no two particles of the same type can occupy precisely the same quantum state. Obeyed by baryons and leptons, but not by photons or mesons.

PHASE TRANSITION   The sharp transition of a system from one configuration to another, usually with a change in symmetry. Examples include melting, boiling, and the transition from ordinary conductivity to superconductivity.

PHOTON   In the quantum theory of radiation, the particle associated with a light wave. Denoted as γ.

PI MESON   The hadron of lowest mass. Comes in three varieties, a positively charged particle (π^+), its negatively charged antiparticle (π^-), and a slightly lighter neutral particle (π^0). Sometimes called *pions*.

PLANCK'S CONSTANT   The fundamental constant of quantum mechanics. Denoted h. Equal to 6.625×10^{-27} erg sec. Planck's constant was first introduced in 1900, in Planck's theory of black-body radiation. It then appeared in Einstein's 1905 theory of photons: the energy of a photon is Planck's constant times the speed of light divided by the wavelength. Today it is more usual to use a constant $\hbar$, defined as Planck's constant divided by 2π.

PLANCK DISTRIBUTION   The distribution of energy at different wavelengths for radiation in thermal equilibrium, i.e., for black-body radiation.

POSITRON   The positively charged antiparticle of the electron. Denoted e^+.

PROPER MOTION   The shift in position in the sky of astronomical bodies, caused by their motion at right angles to the line of sight. Usually measured in seconds of arc per year.

PROTON   The positively charged particle found along with neutrons in ordinary atomic nuclei. Denoted p. The nucleus of hydrogen consists of one proton.

QUANTUM MECHANICS   The fundamental physical theory developed in the 1920s as a replacement for classical mechanics. In quantum mechanics waves and particles are two aspects of the same underlying entity. The particle associated with a given wave is its *quantum*. Also, the states of bound systems like atoms or molecules occupy only certain distinct energy levels; the energy is said to be *quantized*.

QUARKS   Hypothetical fundamental particles, of which all hadrons are supposed to be composed. Isolated quarks have never been observed, and there are theoretical reasons to suspect that, though in some sense real, quarks never *can* be observed as isolated particles.

QUASI-STELLAR OBJECTS A class of astronomical objects with a stellar appearance and very small angular size, but with large red shifts. Sometimes called *quasars*, or when they are strong radio sources, *quasistellar sources*. Their true nature is unknown.

RAYLEIGH-JEANS LAW The simple relation between energy density (per unit wavelength interval) and wavelength, valid for the long-wavelength limit of the Planck distribution. The energy density in this limit is proportional to the inverse fourth power of the wavelength.

RECOMBINATION The combination of atomic nuclei and electrons into ordinary atoms. In cosmology, recombination often is used specifically to refer to the formation of helium and hydrogen atoms at a temperature around 3,000° K.

RED SHIFT The shift of spectral lines toward longer wavelengths, caused by the Doppler effect for a receding source. In cosmology, refers to the observed shift of spectral lines of distant astronomical bodies toward long wavelengths. Expressed as a fractional increase in wavelength, the red shift is denoted z.

REST ENERGY The energy of a particle at rest, which would be released if all the mass of the particle could be annihilated. Given by Einstein's formula $E = mc^2$.

RHO MESON One of many extremely unstable hadrons. Decays into two pi mesons, with a mean life of 4.4×10^{-24} seconds.

SPECIAL RELATIVITY The new view of space and time presented by Albert Einstein in 1905. As in Newtonian mechanics, there is a set of mathematical transformations which relate the space-time coordinates used by different observers, in such a way that the laws of nature appear the same to these observers. However, in special relativity the space-time transformations have the essential property of leaving the speed of light unchanged, irrespective of the velocity of the observer. Any system containing particles with velocities near the speed of light is said to be relativistic, and must be treated according to the rules of special relativity, rather than Newtonian mechanics.

SPEED OF LIGHT The fundamental constant of special relativity, equal to 299,729 kilometers per second. Denoted c. Any particles of zero mass, such as photons, neutrinos, or gravitons, travel at the speed of light. Material particles approach the speed of light when their energies are very large compared to the rest energy mc^2 in their mass.

SPIN A fundamental property of elementary particles which describes the state of rotation of the particle. According to the rules of quantum mechanics, the spin can take only certain special values, equal to a whole number or half a whole number × Planck's constant.

STEADY-STATE THEORY The cosmological theory developed by Bondi, Gold, and Hoyle, in which the average properties of the universe never change with time; new matter must be continually created to keep the density constant as the universe expands.

STEFAN-BOLTZMANN LAW The relation of proportionality between the energy density in black-body radiation and the fourth power of the temperature.

STRONG INTERACTIONS The strongest of the four general classes of elementary particle interactions. Responsible for the nuclear forces which hold protons and neutrons in the atomic nucleus. Strong interactions affect only hadrons, not leptons or photons.

SUPERNOVAS Enormous stellar explosions in which all but the inner core of a star is blown off into interstellar space. A supernova produces in a few days as much energy as the sun radiates in a thousand million years. The last supernova observed in our galaxy was seen by Kepler (and by Korean and Chinese court astronomers) in 1604 in the constellation Ophiuchus, but the radio source Cas A is believed to be due to a more recent supernova.

THERMAL EQUILIBRIUM A state in which the rates at which particles enter any given range of velocities, spins, and so on, exactly balances the rates at which they leave. If left undisturbed for a sufficiently long time, any physical system will eventually approach a state of thermal equilibrium.

THRESHOLD TEMPERATURE The temperature above which a given type of particle will be copiously produced by black-body radiation. Equal to the mass of the particle, times the square of the speed of light, divided by Boltzmann's constant.

TRITIUM The unstable heavy isotope H^3 of hydrogen. Nuclei of tritium consist of one proton and two neutrons.

TYPICAL GALAXIES Here used to refer to galaxies which have no peculiar velocity, and therefore move only with the general flow of matter produced by the expansion of the universe. The same meaning is given here to *typical particle* or *typical observer*.

ULTRAVIOLET RADIATION Electromagnetic waves with wavelength in the range 10 Angstroms to 2,000 Angstroms (10^{-7} cm to 2×10^{-5}cm), intermediate between visible light and X rays.

VIRGO CLUSTER A giant cluster of over 1,000 galaxies in the constellation Virgo. This cluster is moving away from us at a speed of approximately

1,000 km/sec, and is believed to be at a distance of 60 million light years.

WAVELENGTH   In any kind of wave, the distance between wave crests. For electromagnetic waves, the wavelength may be defined as the distance between points where any component of the electric or magnetic field vector takes its maximum value. Denoted λ.

WEAK INTERACTIONS   One of the four general classes of elementary particle interactions. At ordinary energies, weak interactions are much weaker than electromagnetic or strong interactions, though very much stronger than gravitation. The weak interactions are responsible for the relatively slow decays of particles like the neutron and muon, and for all reactions involving neutrinos. It is now widely believed that the weak, electromagnetic and perhaps the strong interactions are manifestations of a simple, underlying unified gauge field theory.

A Mathematical Supplement

These notes are provided for readers who wish to see some of the mathematics that underlie the non-mathematical exposition presented in the body of this book. It should not be necessary to study these notes in order to follow the discussions in the main part of this book.

Note 1   The Doppler Effect

Suppose that wave crests leave a light source at regular intervals separated by a period T. If the source is moving at a velocity V away from the observer, then during the time between successive crests the source moves a distance VT. This increases the time required for the wave crest to get from the source to the observer by an amount VT/c, where c is the speed of light. Thus, the time between arrival of successive wave crests at the observer is

$$T' = T + \frac{VT}{c}$$

The wavelength of the light upon emission is

$$\lambda = cT$$

and the wavelength when the light arrives is

$$\lambda' = cT'$$

Thus, the ratio of these wavelengths is

$$\lambda'/\lambda = T'/T = 1 + \frac{V}{c}$$

The same reasoning applies if the source is moving toward the observer, except that V is replaced with $-V$. (It also applies for any kind of wave signal, not just light waves.)

For instance, the galaxies of the Virgo cluster are moving away from our galaxy at a speed of about 1,000 kilometers per second. The speed of light is 300,000 kilometers per second. Therefore the wavelength λ' of any spectral line from the Virgo cluster is larger than its normal value λ by a ratio

$$\lambda'/\lambda = 1 + \frac{1,000 \text{ km/sec}}{300,000 \text{ km/sec}} = 1.0033$$

166

A Mathematical Supplement

Note 2   The Critical Density

Consider a sphere of galaxies of radius R. (For the purposes of this calculation we must take R to be larger than the distance between clusters of galaxies, but smaller than any distance characterizing the universe as a whole.) The mass of this sphere is its volume times the cosmic mass density ρ:

$$M = \frac{4\pi R^3}{3} \rho$$

Newton's theory of gravitation gives the potential energy of any typical galaxy at the surface of this sphere as

$$P.E. = -\frac{mMG}{R} = -\frac{4\pi m R^2 \rho G}{3}$$

where m is the mass of the galaxy, and G is Newton's constant of gravitation

$$G = 6.67 \times 10^{-8} \text{ cm}^3/\text{gm sec}^2$$

The velocity of this galaxy is given by the Hubble law as

$$V = HR$$

where H is Hubble's constant. Thus its kinetic energy is given by

$$K.E. = \frac{1}{2} mV^2 = \frac{1}{2} mH^2 R^2$$

The total energy of the galaxy is the sum of the kinetic and potential energies

$$E = P.E. + K.E. = mR^2 \left[\frac{1}{2} H^2 - \frac{4}{3} \pi \rho G \right]$$

This quantity must remain constant as the universe expands.

If E is negative the galaxy can never escape to infinity, because at very great distances the potential energy becomes negligible, in which case the total energy is just the kinetic energy, which is always positive. On the other hand, if E is positive the galaxy can reach infinity with some kinetic energy left. Thus, the condition for the galaxy to have just barely escape velocity is that E vanish, which gives

$$\frac{1}{2} H^2 = \frac{4}{3} \pi \rho G$$

In other words, the density must take the value

A Mathematical Supplement

$$\rho_c = \frac{3H^2}{8\pi G}$$

This is the critical density. (Although this result has been derived here using Newtonian physical principles, it is actually valid even when the contents of the universe are highly relativistic, provided that ρ is interpreted as the total energy density divided by c^2.)

For instance, if H has the currently popular value of 15 kilometers per second per million light years, then, recalling that there are 9.46×10^{12} kilometers in a light year, we have

$$\rho_c = \frac{3}{8\pi(6.67 \times 10^{-8} \text{cm}^3/\text{gm sec}^2)} \left(\frac{15 \text{km/sec}/10^6 \text{lt yrs}}{9.46 \times 10^{12} \text{ km/lt yr}}\right)^2$$

$$= 4.5 \times 10^{-30} \text{ gm/cm}^3$$

There are 6.02×10^{23} nuclear particles per gram, so this value for the present critical density corresponds to about 2.7×10^{-6} nuclear particles per cm^3, or 0.0027 particles per liter.

Note 3   Expansion Time Scales

Now consider how the parameters of the universe change with time. Suppose that at a time t a typical galaxy of mass m is at a distance $R(t)$ from some arbitrarily chosen central galaxy, say our own. We saw in the last mathematical note that the total (kinetic plus potential) energy of this galaxy is

$$E = mR^2(t)\left[\frac{1}{2}H^2(t) - \frac{4}{3}\pi\rho(t)G\right]$$

where $H(t)$ and $\rho(t)$ are the values of the Hubble "constant" and the cosmic mass density at time t. This must be a true constant. However, we will see below that $\rho(t)$ increases as $R(t) \to 0$ at least as fast as $1/R^3(t)$, so $\rho(t)R^2(t)$ grows at least as fast as $1/R(t)$ for $R(t)$ going to zero. In order to keep the energy E constant, the two terms in the brackets must therefore nearly cancel, so that for $R(t) \to 0$ we have

$$\frac{1}{2}H^2(t) \to \frac{4}{3}\pi\rho(t)G$$

The characteristic expansion time is just the reciprocal of the Hubble constant, or

A Mathematical Supplement

$$t_{exp}(t) \equiv \frac{1}{H(t)} = \sqrt{\frac{3}{8\pi\rho(t)G}}$$

For instance, at the time of the first frame in Chapter V the mass density was 3.8 thousand million grams per cubic centimeter. Thus, the expansion time then was

$$t_{exp} = \sqrt{\frac{3}{8\pi(3.8 \times 10^9 \text{ gm/cm}^3)(6.67 \times 10^{-8} \text{ cm}^3/\text{gm sec}^2)}} = 0.022 \text{ seconds}$$

Now, how does $\rho(t)$ vary with $R(t)$? If the mass density is dominated by the masses of nuclear particles (the matter-dominated era), then the total mass within a comoving sphere of radius $R(t)$ is just proportional to the number of nuclear particles within that sphere, and hence must remain constant:

$$\frac{4\pi}{3} \rho(t)R(t)^3 = \text{constant}$$

Hence $\rho(t)$ is inversely proportional to $R(t)^3$

$$\rho(t) \propto 1/R(t)^3$$

(The symbol $\propto$ means "is proportional to . . .") On the other hand, if the mass density is dominated by the mass equivalent to the energy of radiation (the radiation-dominated era), then $\rho(t)$ is proportional to the fourth power of the temperature. But the temperature varies like $1/R(t)$, so $\rho(t)$ is then inversely proportional to $R(t)^4$

$$\rho(t) \propto 1/R(t)^4$$

In order to be able simultaneously to deal with the matter- and radiation-dominated eras, we will write these results in the form

$$\rho(t) \propto [1/R(t)]^n$$

with

$$n = \begin{cases} 3 & \text{matter-dominated era} \\ 4 & \text{radiation-dominated era} \end{cases}$$

Note incidentally that $\rho(t)$ does blow up at least as fast as $1/R(t)^3$ for $R(t) \to 0$, as promised.

The Hubble constant is proportional to $\sqrt{\rho}$, and therefore

$$H(t) \propto [1/R(t)]^{n/2}$$

But the velocity of the typical galaxy is then

A Mathematical Supplement

$$V(t) = H(t)R(t) \propto [R(t)]^{1-n/2}$$

It is an elementary result of differential calculus that, whenever the velocity is proportional to some power of the distance, then the time that it takes to go from one point to another is proportional to the change in the ratio of distance to velocity. To be more specific, for V proportional to $R^{1-n/2}$, this relation is

$$t_1 - t_2 = \frac{2}{n}\left[\frac{R(t_1)}{V(t_1)} - \frac{R(t_2)}{V(t_2)}\right]$$

or

$$t_1 - t_2 = \frac{2}{n}\left[\frac{1}{H(t_1)} - \frac{1}{H(t_2)}\right]$$

We can express $H(t)$ in terms of $\rho(t)$, and find that

$$t_1 - t_2 = \frac{2}{n}\sqrt{\frac{3}{8\pi G}}\left[\frac{1}{\sqrt{\rho(t_1)}} - \frac{1}{\sqrt{\rho(t_2)}}\right]$$

Thus, whatever the value of n, the time elapsed is proportional to the change in the inverse square root of the density.

For instance, during the whole of the radiation-dominated era after the annihilation of electrons and positrons, the energy density was given by

$$\rho = 1.22 \times 10^{-35}[T(^\circ K)]^4 \text{ gm/cm}^3$$

(See mathematical note 6, p. 176.) Also, here we have $n = 4$. Thus, the time required for the universe to cool from 100 million degrees to 10 million degrees was

$$t = \frac{1}{2}\sqrt{\frac{3}{8\pi(6.67 \times 10^{-8} \text{ cm}^3/\text{gm sec})}}$$

$$\times\left[\frac{1}{\sqrt{1.22 \times 10^{-35} \times 10^{28} \text{ gm/cm}^3}} - \frac{1}{\sqrt{1.22 \times 10^{-35} \times 10^{32} \text{ gm/cm}^3}}\right]$$

$$= 1.90 \times 10^6 \text{ sec} = 0.06 \text{ years}$$

Our general result can also be expressed more simply by saying that the time required for the density to drop to a value ρ from some value very much greater than ρ is

$$t = \frac{2}{n}\sqrt{\frac{3}{8\pi G\rho}} = \begin{cases} \tfrac{1}{2}\, t_{\mathrm{exp}} & \text{radiation-dominated} \\ \tfrac{2}{3}\, t_{\mathrm{exp}} & \text{matter-dominated} \end{cases}$$

(If $\rho(t_2) \gg \rho(t_1)$), we can neglect the second term in our formula for $t_1 - t_2$.) For instance, at 3,000° K the mass density of photons and neutrinos was

$$\rho = 1.22 \times 10^{-35} \times [3,000]^4 \text{ gm/cm}^3 = 9.9 \times 10^{-22} \text{ gm/cm}^3$$

This is so much less than the density at 10^8 ° K (or 10^7 ° K, or 10^6 ° K) that the time required for the universe to cool from very high early temperatures to 3,000° K may be calculated (setting $n = 4$) simply as

$$\frac{1}{2}\sqrt{\frac{3}{8\pi(6.67 \times 10^{-8} \text{ cm}^3/\text{gm sec}^2)(9.9 \times 10^{-22} \text{ gm/cm}^3)}}$$
$$= 2.1 \times 10^{13} \text{ sec} = 680{,}000 \text{ years}$$

We have shown that the time required for the density of the universe to drop to a value ρ from much higher earlier values is proportional to $1/\sqrt{\rho}$, while the density ρ is proportional to $1/R^n$. The time is therefore proportional to $R^{n/2}$, or, in other words

$$R \propto t^{2/n} = \begin{cases} t^{1/2} & \text{matter-dominated era} \\ t^{2/3} & \text{radiation-dominated era} \end{cases}$$

This remains valid until the kinetic and potential energies have both decreased so much that they are beginning to be comparable to their sum, the total energy.

As remarked in Chapter II, there is at any time t after the beginning a horizon at a distance of order ct, from beyond which no information could yet have reached us. We now see that $R(t)$ vanishes less rapidly as $t \to 0$ than the distance to the horizon, so that at a sufficiently early time any given "typical" particle is beyond the horizon.

Note 4   Black-body Radiation

The *Planck distribution* gives the energy du of black-body radiation per unit volume, in a narrow range of wavelengths from λ to $\lambda + d\lambda$, as

$$du = \frac{8\pi hc}{\lambda^5} d\lambda \Big/ [e^{\left(\frac{hc}{kT\lambda}\right)} - 1]$$

Here T is the temperature; k is Boltzmann's constant (1.38×10^{-16} erg/° K); c is the speed of light (299,729 km/sec); e is the numerical constant 2.718

A *Mathematical Supplement*

. . . ; and h is the Planck constant (6.625×10^{-27} erg sec), originally introduced by Max Planck as an ingredient in this formula.

For *long* wavelengths, the denominator in the Planck distribution may be approximated by

$$e^{\left(\frac{hc}{kT\lambda}\right)} - 1 \simeq \left(\frac{hc}{kT\lambda}\right)$$

Thus in this wavelength region, the Planck distribution gives

$$du = \frac{8\pi kT}{\lambda^4} d\lambda$$

This is the *Rayleigh-Jeans formula*. If this formula held down to arbitrarily small wavelengths, $du/d\lambda$ would become infinite for $\lambda \to 0$, and the total energy density in black-body radiation would be infinite.

Fortunately, the Planck formula for du reaches a maximum at a wavelength

$$\lambda = .2014052 \; hc/kT$$

and then falls steeply off for decreasing wavelengths. The total energy density in the black-body radiation is the integral

$$u = \int_0^\infty \frac{8\pi hc}{\lambda^5} d\lambda \Big/ \left(e^{\left(\frac{hc}{kT\lambda}\right)} - 1\right)$$

Integrals of this sort can be looked up in standard tables of definite integrals; the result is

$$u = \frac{8\pi^5 (kT)^4}{15(hc)^3} = 7.56464 \times 10^{-15} \; [T(^\circ \text{ K})]^4 \text{erg/cm}^3$$

This is the *Stefan-Boltzmann law*.

We can easily interpret the Planck distribution in terms of quanta of light, or photons. Each photon has an energy given by the formula

$$E = hc/\lambda$$

Hence the number dN of photons per unit volume in black-body radiation in a narrow range of wavelengths from λ to $\lambda + d\lambda$ is

$$dN = \frac{du}{hc/\lambda} = \frac{8\pi}{\lambda^4} d\lambda \Big/ [e^{\left(\frac{hc}{kT\lambda}\right)} - 1]$$

The total *number* of photons per unit volume is then

$$N = \int_0^\infty dN = 60.42198 \left(\frac{kT}{hc}\right)^3 = 20.28[T(^\circ \text{ K})]^3 \text{ photons/cm}^3$$

and the average photon energy is

$$E_{\text{average}} = u/N = 3.73 \times 10^{-16}[T(^\circ \text{ K})] \text{ ergs}$$

Now let's consider what happens to black-body radiation in an expanding universe. Suppose the size of the universe changes by a factor f; for instance, if it doubles in size, then $f = 2$. As we saw in Chapter II, the wavelengths will change in proportion to the size of the universe to a new value

$$\lambda' = f\lambda$$

After the expansion, the energy density du' in the new wavelength range λ' to $\lambda' + d\lambda'$ is less than the original energy density du in the old wavelength range λ to $\lambda + d\lambda$, for two different reasons:

1. Since the volume of the universe has increased by a factor f^3, as long as no photons have been created or destroyed, the number of photons per unit volume has decreased by a factor $1/f^3$.

2. The energy of each photon is inversely proportional to its wavelength, and is therefore decreased by a factor $1/f$. It follows that the energy density is decreased by an overall factor $1/f^3$ times $1/f$, or $1/f^4$:

$$du' = \frac{1}{f^4} du = \frac{8\pi hc}{\lambda^5 f^4} d\lambda \Big/ [e^{\left(\frac{hc}{kT\lambda}\right)} - 1]$$

If we rewrite this formula in terms of the new wavelengths λ', it becomes

$$du' = \frac{8\pi hc}{\lambda'^5} d\lambda' \Big/ [e^{\left(\frac{hcf}{kT\lambda'}\right)} - 1]$$

But this is exactly the same as the old formula for du in terms of λ and $d\lambda$, except that T has been replaced with a new temperature

$$T' = T/f$$

Thus, we conclude that freely expanding black-body radiation remains described by the Planck formula, but with a temperature that drops in inverse proportion to the scale of the expansion.

Note 5   The Jeans Mass

In order for a clump of matter to form a gravitationally bound system, it is necessary for its gravitational potential energy to exceed its internal thermal

A Mathematical Supplement

energy. The gravitational potential energy of a clump of radius r and mass M is of order

$$P.E. \approx -\frac{GM^2}{r}$$

The internal energy per unit volume is proportional to the pressure p, so the total internal energy is of order

$$I.E. \approx pr^3$$

Thus gravitational clumping should be favored if

$$\frac{GM^2}{r} \gg pr^3$$

But for a given density ρ we can express r in terms of M through the relation

$$M = \frac{4\pi}{3}\rho r^3$$

The condition for gravitational clumping may therefore be written

$$GM^2 \gg p(M/\rho)^{4/3}$$

or in other words

$$M \gg M_J$$

where M_J is (within an inessential numerical factor) the quantity known as the *Jeans mass*:

$$M_J = \frac{p^{3/2}}{G^{3/2}\rho^2}$$

For instance, just before the recombination of hydrogen, the mass density was 9.9×10^{-22} gm/cm³ (see mathematical note 3, p. 171), and the pressure was

$$p \approx \frac{1}{3}c^2\,\rho = 0.3 \text{ gm/cm sec}^2$$

The Jeans mass was therefore

$$M_J = \left(\frac{0.3 \text{ gm/cm sec}^2}{6.67 \times 10^{-8} \text{ cm}^3/\text{gm sec}^2}\right)^{3/2}\left(\frac{1}{9.9 \times 10^{-22} \text{ gm/cm}^3}\right)^2$$
$$= 9.7 \times 10^{51} \text{ gm} = 5 \times 10^{18}\,M_\odot$$

where $M_\odot$ is one solar mass. (In comparison, the mass of our galaxy is about $10^{11} M_\odot$.) After recombination, the pressure dropped by a factor 10^9, so the Jeans mass dropped to

$$M_J = (10^{-9})^{3/2} \times 5 \times 10^{18}\ M_\odot = 1.6 \times 10^5 M_\odot$$

It is interesting that this is roughly the mass of the large globular clusters within our galaxy.

Note 6   Neutrino Temperature and Density

As long as thermal equilibrium is preserved, the total value of the quantity known as "entropy" remains fixed. For our purposes, the entropy per unit volume S is given to an adequate approximation at temperature T by

$$S \propto N_T T^3$$

where N_T is the effective number of species of particles in thermal equilibrium whose threshold temperature lies below T. In order to keep the total entropy constant, S must be proportional to the inverse cube of the size of the universe. That is, if R is the separation between any pair of typical particles, then

$$SR^3 \propto N_T T^3 R^3 = \text{constant}$$

Just before the annihilation of electrons and positrons (at about $5 \times 10^9\ ^\circ$ K) the neutrinos and antineutrinos had already gone out of thermal equilibrium with the rest of the universe, so the only abundant particles in equilibrium were the electron, positron, and photon. Referring to Table One on page 156, we see the effective total number of particle species before annihilation was

$$N_{\text{before}} = \frac{7}{2} + 2 = \frac{11}{2}$$

On the other hand, after the annihilation of electrons and positrons in the fourth frame, the only remaining abundant particles in equilibrium were the photons. The effective number of particle species then was simply

$$N_{\text{after}} = 2$$

It follows then from the conservation of entropy that

$$\frac{11}{2}(TR)^3_{\text{before}} = 2(TR)^3_{\text{after}}$$

A Mathematical Supplement

That is, the heat produced by the annihilation of electrons and positrons increases the quantity TR by a factor

$$\frac{(TR)_{\text{after}}}{(TR)_{\text{before}}} = \left(\frac{11}{4}\right)^{1/3} = 1.401$$

Before the annihilation of electrons and positrons, the neutrino temperature T_ν was the same as the photon temperature T. But from then on, T_ν simply dropped like $1/R$, so for all subsequent times $T_\nu R$ was equal to the value of TR before the annihilation:

$$(T_\nu R)_{\text{after}} = (T_\nu R)_{\text{before}} = (TR)_{\text{before}}$$

We conclude therefore that after the annihilation process is over, the photon temperature is higher than the neutrino temperature by a factor

$$(T/T_\nu)_{\text{after}} = \frac{(TR)_{\text{after}}}{(T_\nu R)_{\text{after}}} = \left(\frac{11}{4}\right)^{1/3} = 1.401$$

Even though out of thermal equilibrium, the neutrinos and antineutrinos make an important contribution to the cosmic energy density. The effective number of species of neutrinos and antineutrinos is 7/2, or 7/4 of the effective number of species of photons. (There are two photon spin states.) On the other hand, the fourth power of the neutrinos' temperature is less than the fourth power of the photon temperature by a factor $(4/11)^{4/3}$. Thus the ratio of the energy density of neutrinos and antineutrinos to that of photons is

$$\frac{u_\nu}{u_\gamma} = \frac{7}{4}\left(\frac{4}{11}\right)^{4/3} = 0.4542$$

The Stefan-Boltzmann law (see Chapter III) tells us that at *photon* temperature T the photon energy density is

$$u_\gamma = 7.5641 \times 10^{-15} \text{ erg/cm}^3 \times [T(^\circ \text{ K})]^4$$

Hence the total energy density after electron-positron annihilation is

$$u = u_\nu + u_\gamma = 1.4542 u_\gamma = 1.100 \times 10^{-14} \text{ erg/cm}^3 [T(^\circ \text{ K})]^4$$

We can convert this to an equivalent mass density by dividing by the square of the speed of light, and find

$$\rho = u/c^2 = 1.22 \times 10^{-35} \text{ gm/cm}^3 \times [T(^\circ \text{ K})]^4$$

Suggestions for Further Reading

A. Cosmology and General Relativity

The following treatises provide an introduction to various aspects of cosmology, and to those parts of general relativity relevant to cosmology, on a level that is generally more technical than that of this book.

Bondi, H. *Cosmology* (Cambridge University Press, Cambridge, England, 1960). By now somewhat out of date, but contains interesting discussions of the Cosmological Principle, steady-state cosmology, Olber's paradox, and so on. Very readable.

Eddington, A. S. *The Mathematical Theory of Relativity*, 2nd ed. (Cambridge University Press, Cambridge, England, 1924). For many years the leading book on general relativity. Historically interesting early discussion of red shifts, de Sitter model, and so on.

Einstein, A., et al. *The Principle of Relativity* (Methuen and Co., Ltd., London, 1923; reprinted by Dover Publications, Inc., New York). Invaluable reprints of original papers on special and general relativity by Einstein, Minkowski, and Weyl, in English translation. Includes reprint of Einstein's 1917 paper on cosmology.

Field, G. B.; Arp, H.; and Bahcall, J. N. *The Redshift Controversy* (W. A. Benjamin, Inc., Reading, Mass., 1973). A remarkable debate on the interpretation of red shifts in terms of a cosmological recession, plus useful reprints of original articles.

Hawking, S. W., and Ellis, G. F. R. *The Large Scale Structure of Space-Time* (Cambridge University Press, Cambridge, England, 1973). Rigorous mathematical treatment of the problem of singularities in cosmology and gravitational collapse.

Hoyle, Fred. *Astronomy and Cosmology—A Modern Course* (W. H. Freeman & Co., San Francisco, 1975). An elementary astronomy textbook, with more of an emphasis on cosmology than usual. Very little mathematics used.

Misner, C. W.; Thorne, K. S.; and Wheeler, J. A. *Gravitation* (W. H.

Freeman & Co., San Francisco, 1973). Up-to-date, comprehensive introduction to general relativity by three leading professionals. Some discussion of cosmology.

O'Hanian, Hans C. *Gravitation and Space Time* (Norton & Company, New York, 1976). A textbook on relativity and cosmology for undergraduates.

Peebles, P. J. E. *Physical Cosmology* (Princeton University Press, Princeton, 1971). Authoritative general introduction, with strong emphasis on observational background.

Sciama, D. W. *Modern Cosmology* (Cambridge University Press, Cambridge, England, 1971). Very readable broad introduction to cosmology and other topics in astrophysics. Written at a level "intelligible to readers with only a modest knowledge of mathematics and physics," with equations held to a minimum.

Segal, I. E. *Mathematical Cosmology and Extragalactic Astronomy* (Academic Press, New York, 1976). For one example of a heterodox but thought-provoking view of modern cosmology.

Tolman, R. C. *Relativity, Thermodynamics, and Cosmology* (Clarendon Press, Oxford, 1934). For many years the standard treatise on cosmology.

Weinberg, Steven. *Gravitation and Cosmology: Principles and Applications of the General Theory of Relativity* (John Wiley & Sons, Inc., New York, 1972). A general introduction to the General Theory of Relativity. About one-third of the volume deals with cosmology. Modesty forbids further comment.

B.   *History of Modern Cosmology*

The following include both firsthand and secondary sources for the history of modern cosmology. Most of these books use little mathematics, but some assume a measure of familiarity with physics and astronomy.

Baade, W. *Evolution of Stars and Galaxies.* (Harvard University Press, Cambridge, Mass., 1968). Lectures given by Baade in 1958, edited from tape recordings by C. Payne-Gaposchkin. Highly personal account of the development of astronomy in this century, including the development of the extragalactic distance scale.

Dickson, F. P. *The Bowl of Night* (M.I.T. Press, Cambridge, Mass., 1968). Cosmology from Thales to Gamow. Contains facsimiles of original articles by de Cheseaux and Olbers, on the darkness of the night sky.

Suggestions for Further Reading

Gamow, George. *The Creation of the Universe* (Viking Press, New York, 1952). Not up to date but valuable as a statement of Gamow's point of view circa 1950. Written for the general public, with Gamow's usual charm.

Hubble, E. *The Realm of the Nebulae* (Yale University Press, New Haven, 1936; reprinted by Dover Publications, Inc., New York, 1958). Hubble's classic account of the astronomical exploration of galaxies, including the discovery of the relation between red shift and distance. Originally delivered as the 1935 Silliman lectures at Yale.

Jones, Kenneth Glyn. *Messier Nebulae and Star Clusters* (American-El-sevier Publishing Co., New York, 1969). Historical notes on the Messier catalog and on the observations of the objects it contains.

Kant, Immanuel. *Universal Natural History and Theory of the Heavens.* Translated by W. Hasties (University of Michigan Press, Ann Arbor, 1969). Kant's famous work on the interpretation of the nebulae as galaxies like our own. Also includes a useful introduction by M. K. Munitz, and a contemporary account of Thomas Wright's theory of the Milky Way.

Koyré, Alexandre. *From the Closed World to the Infinite Universe* (Johns Hopkins Press, Baltimore, 1957; reprinted by Harper & Row, New York, 1957). Cosmology from Nicholas of Cusa to Newton. Contains interesting account of the Newton-Bentley correspondence concerning absolute space and the origin of stars, including useful excerpts.

North, J. D. *The Measure of the Universe* (Clarendon Press, Oxford, 1965). Cosmology from the nineteenth century to the 1940s. Very detailed account of the beginnings of relativistic cosmology.

Reines, F., ed. *Cosmology, Fusion, and Other Matters: George Gamow Memorial Volume* (Colorado Associated University Press, 1972). Valuable firsthand account by Penzias of the discovery of the microwave background, and by Alpher and Herman of the development of the "big bang" model of nucleosynthesis.

Schlipp, P. A., ed. *Albert Einstein: Philosopher-Scientist* (Library of Living Philosophers, Inc., 1951; reprinted by Harper & Row, New York, 1959). Volume 2 contains articles by Lemaitre on Einstein's introduction of the "cosmological constant," and by Infeld on relativistic cosmology.

Shapley, H., ed. *Source Book in Astronomy 1900–1950* (Harvard University Press, Cambridge, Mass., 1960). Reprints of original articles on cosmology and other areas of astronomy, many unfortunately abridged.

Suggestions for Further Reading

C. Elementary Particle Physics

There are as yet no books that deal on a nonmathematical level with most of the recent developments in elementary particle physics discussed in Chapter VII. The following article should provide an introduction of sorts:

Weinberg, Steven, "Unified Theories of Elementary Particle Interaction," *Scientific American*, July 1974, pp. 50–59.

For a more comprehensive introduction to elementary particle physics that is soon to be published, see: Feinberg, G. *What is the World Made of? The Achievements of Twentieth Century Physics* (Garden City: Anchor Press/Doubleday, 1977).

For an introduction written for specialists, with references to the original literature, see either of the following:

Taylor, J. C. *Gauge Theories of Weak Interactions* (Cambridge University Press, Cambridge, England, 1976).
Weinberg, S. "Recent Progress in Gauge Theories of the Weak, Electromagnetic, and Strong Interactions," *Reviews of Modern Physics*, Vol. 46, pp. 255–277 (1974).

D. Miscellaneous

Allen, C. W. *Astrophysical Quantities*. 3rd ed. (The Athlone Press, London, 1973). A handy collection of astrophysical data and formulas.
Sandage, A. *The Hubble Atlas of Galaxies* (Carnegie Institute of Washington, Washington, D.C., 1961). A large number of beautiful photographs of galaxies, assembled to illustrate the Hubble classification scheme.
Sturleson, Snorri. *The Younger Edda*, translated by R. B. Anderson (Scott, Foresman & Co., Chicago, 1901). For another view of the beginning and end of the universe.

Index

Index

Index

Index

Photons (continued)
54; energies of, 61-62; in first three minutes of universe, 103-109; in first thirty-four minutes of universe, 112; in first 700,000 years of universe, 54-55; lepton number per, 99, 100; mean free time of, 55; other particles and, 54, 66, 82, 84; in quantum theory, 53, 80; in quark theory, 139; as radiation, 57, 60; in radiation dominated era, 78, 79, 81; spin of, 85; wavelength, 62-65, 69, 76

Photon-nuclear particle ratio, 74, 113-116, 120, 126, 129, 131; consequences of, 75-76; cosmic deuterium abundance and, 114-116; in present universe, 73, 95, 120; in radiation dominated era, 78; in steady-state theory, 154; table of possible values, 114

Photosynthesis, 62

Planck, Max Karl Ernst Ludwig, 58, 60

Planck's constant, 162

Planck distribution, 58-61, 67, 69, 70, 162

Planck black body formula, 64-66

Pleiades (M45), 17

Plutonium, 62

Polar star, 11

Politzer, Hugh David, 140

Positrons, 83-85, 87, 92, 102, 124, 152, 162; discovery of, 83, 127; e⁺, 81, 85; in first 1/100 second of universe, 103, 105; in first three minutes of universe, 6, 7, 104-109; at first 34 minutes of universe, 112

Proceedings of the Royal Danish Academy, 125

Proper motion, defined, 12, 162; as measure of distance, 16, 25

Protons, 83, 91, 133, 162; as baryons, 92-93, 95; in contracting universe, 152; density of, 73; energy of, 74; after first few minutes of the universe, 87; in first three minutes of the universe, 6, 7, 104-107, 109, 110; in first 34 minutes of the universe, 112; helium nuclei and, 128; in quark model, 141; ratio to photons, 74, 75; strong interaction and, 134, 135; See also Neutron-proton balance

Quanta, 60, 61, 78

Quantum mechanics, 60, 93, 153, 162

Quantum theory, 6, 53, 57, 78, 80, 141

Quark theory, 139-142, 162

Quasi-stellar objects, 29, 117, 162

Radiation: defined, 50; background, 50-51, 66-68, 72 (See also Cosmic microwave radiation background); in early universe, 50-51; quantum view of, 53; in thermal equilibrium, 57; See also Black body radiation

Radiation-dominated era, 76, 78, 79

Radiation pressure, 74-75

Radio astronomy, 45, 67-69, 130

Radio noise, 46, 48, 49, 51

Rayleigh, Lord, 67

Rayleigh-Jeans distribution, 58

Rayleigh-Jeans Law, 163

Rayleigh-Jeans region, 67

Recombination, 163

Red shift: of Capella, 15; defined, 163; in contracting universe, 151; in de Sitter's model, 33; differing interpretations of, 28-29; distance curve, 38, 39; effect of, 65; in Einstein's model, 32; equivalent temperature of radiation background and, 50; of galaxies, 21; light waves and, 30-31; neutrino wave lengths and, 107; photon wave length and, 78; of quasi-stellar objects, 29

Reines, F., 179

Rest energy, 80-83, 163

Relativity: General Theory of, 25, 32, 35, 37; Special Theory of, 25, 80, 83

Rho mesons, 138, 141, 163

Roll, P. G., 52, 53, 66, 153

Rotating state, of cyanogen molecule, 69

Rutherford, Lord, 30

Salam, Abdus, 143

Salpeter, E. E., 128

Sandage, Allan, 27, 38, 180

Satellites, artificial earth, 70, 72; *Copernicus*, 115-117; *Echo*, 45

Saturn, 15

Schlipp, P. A., 179

186

Index

Index